Screen Ecologies

Leonardo

Roger F. Malina, Executive Editor

Sean Cubitt, Editor-in-Chief

The Hidden Sense: Synesthesia in Art and Science, Cretien van Campen, 2007

Closer: Performance, Technologies, Phenomenology, Susan Kozel, 2007

Video: The Reflexive Medium, Yvonne Spielmann, 2007

Software Studies: A Lexicon, Matthew Fuller, 2008

Tactical Biopolitics: Art, Activism, and Technoscience, edited by Beatriz da Costa and Kavita Philip, 2008

White Heat and Cold Logic: British Computer Art, 1960–1980, edited by Paul Brown, Charlie Gere, Nicholas Lambert, and Catherine Mason, 2008

Rethinking Curating: Art after New Media, Beryl Graham and Sarah Cook, 2010

Green Light: Toward an Art of Evolution, George Gessert, 2010

Enfoldment and Infinity: An Islamic Genealogy of New Media Art, Laura U. Marks, 2010

Synthetics: Aspects of Art and Technology in Australia, 1956–1975, Stephen Jones, 2011

Hybrid Cultures: Japanese Media Arts in Dialogue with the West, Yvonne Spielmann, 2012

Walking and Mapping: Artists as Cartographers, Karen O'Rourke, 2013

The Fourth Dimension and Non-Euclidean Geometry in Modern Art, revised edition, Linda Dalrymple Henderson, 2013

Illusions in Motion: Media Archaeology of the Moving Panorama and Related Spectacles, Erkki Huhtamo, 2013

Relive: Media Art Histories, edited by Sean Cubitt and Paul Thomas, 2013

Re-collection: Art, New Media, and Social Memory, Richard Rinehart and Jon Ippolito, 2014

Biopolitical Screens: Image, Power, and the Neoliberal Brain, Pasi Väliaho, 2014

The Practice of Light: A Genealogy of Visual Technologies from Prints to Pixels, Sean Cubitt, 2014

The Tone of Our Times: Sound, Sense, Economy, and Ecology, Frances Dyson, 2014

The Experience Machine: Stan VanDerBeek's Movie-Drome and Expanded Cinema, Gloria Sutton, 2014

Hanan al-Cinema: Affections for the Moving Image, Laura U. Marks, 2015

Writing and Unwriting (Media) Art History: Erkki Kurenniemi in 2048, edited by Joasia Krysa and Jussi Parikka, 2015

Control: Digitality as Cultural Logic, Seb Franklin, 2015

New Tendencies: Art at the Threshold of the Information Revolution (1961–1978), Armin Medosch, 2016

Screen Ecologies: Art, Media, and the Environment in the Asia-Pacific Region; Larissa Hjorth, Sarah Pink, Kristen Sharp, and Linda Williams, 2016

See http://mitpress.mit.edu for a complete list of titles in this series.

Screen Ecologies

Art, Media, and the Environment in the Asia-Pacific Region

Larissa Hjorth, Sarah Pink, Kristen Sharp, and Linda Williams

The MIT Press
Cambridge, Massachusetts
London, England

This book was set in Stone Sans and Stone Serif by Toppan Best-set Premedia Limited. Printed and bound in the United States of America.

Library of Congress Cataloging-in-Publication Data

Names: Hjorth, Larissa, author. | Pink, Sarah, author. | Sharp, Kristen, author. | Williams, Linda, 1946– author.
Title: Screen ecologies : art, media, and the environment in the Asia-Pacific region / Larissa Hjorth, Sarah Pink, Kristen Sharp, and Linda Williams.
Description: Cambridge, MA : The MIT Press, 2016. | Series: Leonardo book series | Includes bibliographical references and index.
Identifiers: LCCN 2015039891 | ISBN 9780262034562 (hardcover : alk. paper)
Subjects: LCSH: Art and society—Pacific Area—History—21st century. | Art and technology—Pacific Area—History—21st century. | Art—Pacific Area—21st century—Themes, motives. | Climatic changes—Social aspects—Pacific Area—History—21st century.
Classification: LCC N72.S6 H56 2016 | DDC 701/.03—dc23 LC record available at http://lccn.loc.gov/2015039891

10   9   8   7   6   5   4   3   2   1

Contents

Series Foreword

Leonardo/International Society for the Arts, Sciences, and Technology (ISAST)

Leonardo, the International Society for the Arts, Sciences, and Technology, and the affiliated French organization Association Leonardo have some very simple goals:

1. To advocate, document, and make known the work of artists, researchers, and scholars developing the new ways that the contemporary arts interact with science, technology, and society.
2. To create a forum and meeting places where artists, scientists, and engineers can meet, exchange ideas, and, when appropriate, collaborate.
3. To contribute, through the interaction of the arts and sciences, to the creation of the new culture that will be needed to transition to a sustainable planetary society.

When the journal *Leonardo* was started some forty-five years ago, these creative disciplines existed in segregated institutional and social networks, a situation dramatized at that time by the "Two Cultures" debates initiated by C. P. Snow. Today we live in a different time of cross-disciplinary ferment, collaboration, and intellectual confrontation enabled by new hybrid organizations, new funding sponsors, and the shared tools of computers and the Internet. Above all, new generations of artist-researchers and researcher-artists are now at work individually and collaboratively bridging the art, science, and technology disciplines. For some of the hard problems in our society, we have no choice but to find new ways to couple the arts and sciences. Perhaps in our lifetime we will see the emergence of "new Leonardos," hybrid creative individuals or teams that will not only develop a meaningful art for our times but also drive new agendas in science and stimulate technological innovation that addresses today's human needs.

For more information on the activities of the Leonardo organizations and networks, please visit our websites at http://www.leonardo.info/ and http://www.olats.org.

Roger F. Malina
Executive Editor, Leonardo Publications

Acknowledgments

The authors would like to acknowledge the Australian Research Council, which funded the research project *Spatial Dialogues: Public Art and Climate Change* (LP100200088); Grocon and Fairfax Media; the Schools of Art and Media and Communication at RMIT University; Fine Arts College of Shanghai University; Shanghai Institute of Visual Arts; the Himalayas Art Museum (Shanghai); Mr Genichi Ide and the BOAT PEOPLE Association (Tokyo); Musashino Art University (Tokyo); Professor Fumitoshi Kato (Tokyo); Kana Ohashi; and Professor Christophe Charles. The authors would also like to thank Doug Sery and Sean Cubitt, as well as the anonymous reviewers, for all their invaluable feedback and support.

Larissa Hjorth would like to thank all the artists and non-artists who participated in *keitai mizu*. Artists include Masato Takasaka, Kate Shaw, Kate Rohde, Toshiaki Tomita, Ryuta Nakajima, Yasuko Toyoshima, Fleur Summers, and Simon Perry. Sarah Pink and Larissa Hjorth would also like to acknowledge the Australian Research Council linkage with Intel, *Locating the Mobile* (LP130100848). A big thank you goes to Esther Pierini, Helen Addison-Smith, Natalie King, Erin Taylor, Jen Rae and Gina Krone.

Linda Williams would like to convey special thanks to Mr. David Waldren, the National Executive Design Manager for Grocon, and also Professor Ling Min of the Fine Arts College of Shanghai University.

Larissa Hjorth would like to dedicate this book to the living memory of her mother, artist Noela Hjorth (1940–2016).

1　An Introduction

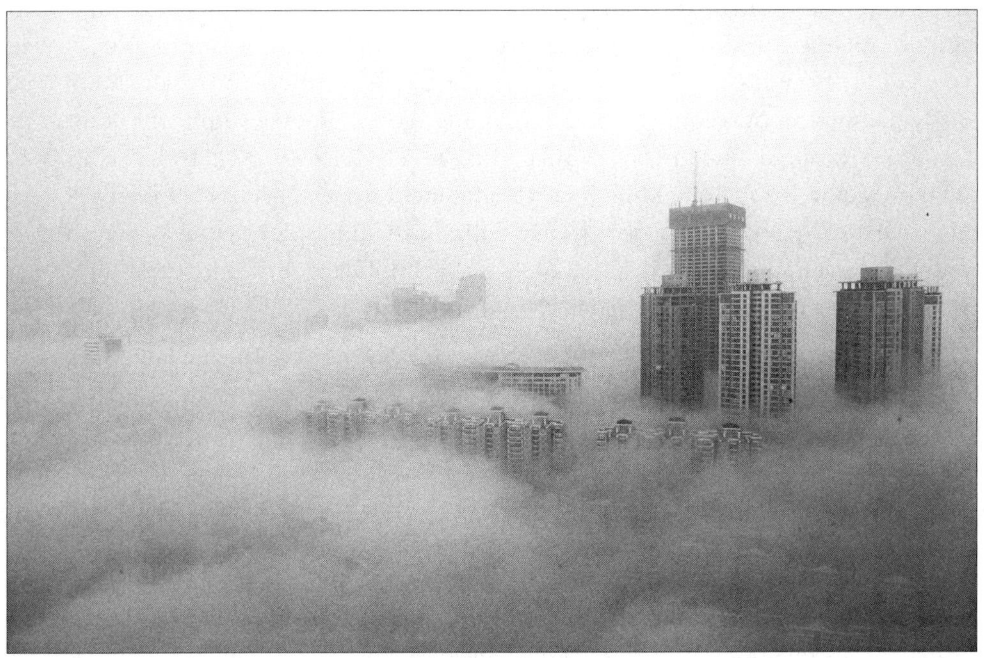

Figure 1.1
Buildings in Lianyungang, China, shrouded by smog, 2013. Photo: Getty Images.

In recent years, a visual culture focusing on environmental deterioration has emerged
as an inescapable element of the way in which social media and mobile technologies
are experienced globally. Accelerated images of environmental disasters, and more generally of pollution and its impact on the environment, have seemingly become part of
our everyday media lives. This visual culture is emerging as part of a wider move toward
recognition, consciousness, and critique of the politics, flows of goods, and capital and

consumer cultures that are beginning to be held responsible for environmental degradation. Reports such as the Intergovernmental Panel on Climate Change (IPCC) Fifth Assessment Report (2014) corroborate with the visual culture that the atmosphere is warming up and the oceans are overheating.

Yet attempts to generate awareness about environmental issues through digital and visual media technologies are riddled with a deep irony. As the anthropologist Thomas Hylland Eriksen's *Overheating* project has shown, an underlying tension exists between environmental sustainability and development (2009, 13). As he puts it, "People want economic growth and a sustainable environment at the same time. Very often you can't have both. That's the fundamental double bind of contemporary global civilization" (n.p.). According to Eriksen, the global warming narrative, born out of the "North Atlantic middle classes," "feeds into an even more comprehensive story about acceleration" (13). It reveals the unintentional consequences of modernity, which are especially manifest in burgeoning Asian developing economies like China and India. This tension is brought out further, with particular relevance to digital media, through Richard Maxwell and Toby Miller's (2012) argument for a "greening" of media studies. Maxwell and Miller's critical perspective outlines the multiple ways in which the production, consumption, and disposal of media technologies and content are implicated in the making of a series of global ills:

Media technologies generate meaning, but also detritus and disease. Their industrial life cycles extend far-flung injuries into natural and bio-physical environments. Their by-products travel the Earth via an international division of labour. (Maxwell and Miller 2012, 165)

As Alison Anderson notes in her discussion of the 2013 IPCC report, "The science linking tropical storms and climate change is unclear" (2014, 1). However, it is easy to imagine how environmental degradation is augmented by globally distributed resource extraction, manufacturing, transportation, and waste—all of which feed or emerge from the processes of development of which the use of social and media technologies is part. It seems similarly reasonable to judge that these could be contributors to the overheating of the oceans, leading to the formation of cyclones and superstorms like Haiyan in the Philippines in 2013 and Sandy in 2012. Indeed, as Anderson goes on to point out, the IPCC report "concludes that the weight of evidence for anthropogenic (human-caused) climate change is mounting" (2014, 1).

Debates about climate change have likewise had their impact in visual culture studies. Nicholas Mirzoeff has argued that what he calls "anthropocene visuality" thrives on crises between "democracy, food supply and climate change" (2014, 230). "Anthropocene" is a term that refers to the idea that the present age is, as the new media scholar and curator Joanna Zylinska puts it, "a geo-historical period, in which humans are said to have become the biggest threat to life on earth" (2014, 10). Mirzoeff argues that we must counter dominant anthropocene visuality through a postcolonial politics that contests the global North (and its exploitation of the global South), creating new ways for thinking about sustainability and the democratic way of life (2014, 230).

As we noted earlier, there is an emerging visual culture around climate change, and in considering this, we should not forget that the IPCC report used digital visualization technologies in making a compelling case for the impact of climate change. However, scientists are not the only ones producing visualizations of media and the environment. An increasing number of artists are merging visual and new media methods to comment on, and intervene in, climate change narratives. This is the focus of this book: how artists help to provide alternative ways in which to understand and visualize the entanglements between media and the environment in the Asia-Pacific. In the following chapters, we consider various debates and artworks. Some, as we show, are creating alternative critical perspectives on the impact of favoring modernization over the environment and are doing so through art practices that go beyond the making of visual culture.

For example, one location that Eriksen (Khazaleh 2013) has commented on is Australia's Great Barrier Reef, which he sees as falling prey to the country's emphasis on economic growth over environmental sustainability. It is against that contradictory backdrop that we can interpret how artists in Australia have begun to focus on representations of, and interventions within, the environment. The Australian artist Kate Shaw associates climate change with processes that are transforming what is considered sublime and familiar. Her landscapes are man-made disasters that are as beautiful as they are toxic. Her challenge to the place of the sublime in art invokes a critique of the relationship between development and climate. Shaw's work breaks down the conventional definition of the sublime as "timeless," which Sean Cubitt points out has tended to persist across shifts in artistic depictions (2005, 46). It likewise deploys a contemporary reality to confront the historical tendency of the sublime in painting to rest in the romanticism of nature (Mirzoeff 2014). As the title of Shaw's work *Contra Natura* suggests, this contemporary reality is destructive of what nature was thought to be.

Screen Ecologies engages with and advances the agenda developed through these scholarly and art critiques of climate change. In doing so, we call for attention to the ways in which critical contemporary digital screen and art cultures are emerging within this complex and ironic context where the creative, generative, and destructive

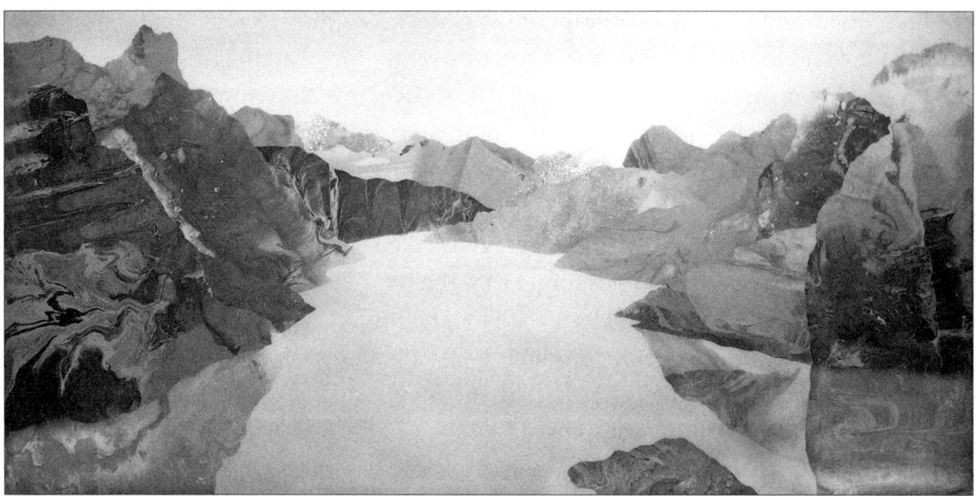

Figure 1.2
Kate Shaw, *Contra Natura*, 2011.

potentials of media have all become part of the same processes and environment. Such an understanding, we argue, provides a starting point for considering how, in the future, digital screen and art cultures might become resilient to the environmentally destructive processes and forces through which they are currently produced, consumed, and sustained.

To achieve this, we move beyond the dependence of media studies on theories from sociology, political science, psychology, and the culturalist stance of visual culture studies. We take an interdisciplinary approach that engages an anthropological theory of environment that situates media (in its production, content, and use) and waste more firmly within the environment. We argue that new knowledge about how digital media are already participating at the interface between critical debate, corporate and political interests, and public engagement is needed. Such knowledge will help us to consider how a world in which the kinds of "green citizenship" that Maxwell and Miller (2012) call for, and where political-economic policies might work against climate change, might be possible. Moreover, to understand this context and the issues it raises, analysis must go beyond the conventional backyard of media studies of Europe and the United States. This means acknowledging the role of the Asia-Pacific region in defining issues, challenges, and potential ways forward. In this region, the manufacture of media technologies dominates, and the consumption and use of digital media technologies and platforms are rising rapidly, making the Asia-Pacific a critical site for analysis.

In *Screen Ecologies* we advance the agenda to better understand and consider future actions in relation to media and climate change by focusing directly on the ways in which screen media technologies, media content, and public art are involved in the relationship between digital media and environmental sustainability. We go beyond an analysis of the textual discourses, political economy, and other objects of study that have been dominant foci in media and cultural studies, to understand how media and arts practices, technologies, and publics participate in these processes.

In the remainder of this chapter, we outline how we conceptualize the Asia-Pacific context in this book. First we ask: Why the Asia-Pacific region? We then set the scene for the subsequent chapters by offering readers a taster of how screen ecologies of the Asia-Pacific region play out. The purpose of this brief description is to demonstrate how relationships between art, screen media, and ecology are being configured in the region, and why they are relevant.

Why the Asia-Pacific? Thinking about the Region

The Asia-Pacific is a region of political, economic, and cultural diversity. This makes it difficult to study as a region, not least because, like any region, it cannot be isolated from other parts of the world. For some, the rubric of "Asia-Pacific" is too problematic to be useful, given that it is laden with vexed geopolitical histories and significant internal variations. These scholars have instead opted to view the region as part of a broader twenty-first-century notion of the "global South" (Dirlik 2007; Ling and Horst 2011). However, a focus on the global South also tends to underplay some aspects of the region's history. Of particular concern for this book is the need to acknowledge processes related to consumption in the region, particularly the ways in which it has participated in the development of information and communications technologies (ICTs) globally (Hjorth 2009; Otmazgin 2014). ICTs have played a key role in the rise of the region, as well as in its transnational collaborations and in the reframing of national boundaries (Berry, Liscutin, and Mackintosh 2009; Otmazgin 2014). This alerts us to the need to attend to the complex entanglements and transnational flows that under-pin the ways the region is defined and articulated.

Part of our rationale for continuing to use "Asia-Pacific" is that the term enables us to map out a particular geography of environmental issues, social media, and public art that in itself operates as a metaphor for reimagining the region—often on the terms of those who are themselves actors in the region, including artists and art networks. It has also been convincingly argued that there is some urgency to attend to questions of climate change in the Asia-Pacific as a region, which is home to the greatest number of megacities (defined by the Asian Development Bank [2012] as cities where the population exceeds 10 million) and is an area that scientific consensus has confirmed is now threatened by significant climatic change and ecological deterioration (CSIRO 2006; IPCC 2011). Nations in the Asia-Pacific are, moreover, "four times more likely to be affected by natural disasters than in Africa, and 25 times more likely than in Europe or North America," and "among the main causes" of this are "unfettered development, urbanization and climate change" (Asian Development Bank 2012, n.p.).

This does not mean that we wish to put a hard-and-fast definition on what the Asia-Pacific stands for. Since its conception after World War II, the region has taken on various contested geographic definitions. For some, it is the Western Pacific Ocean. For others, it is Asia, Australasia, and Oceania. In the late 1980s and early 1990s, the rubric began to gain more currency as the region itself unevenly began to develop global "soft power" through its economic, and thus political, growth. Much of this growth in locations such as Japan and South Korea occurred in and around production by ICTs for global consumption. Institutions such as Asia-Pacific Economic Cooperation (APEC) have played a key role in the last few decades to harness definitions and infrastructure by and for the region (rather than being defined by "others" such as the United States). Transnational migration and the mobility of people, commodities, technologies, and finance across the region have also operated to challenge the rubric's stability, as well as simultaneously creating enclaves and boundaries (Turner 2007).

The composition of the Asia-Pacific when drawn together in practice, however, tends to be contextual and not to function as a group or consortium as a whole. Rather, the term is used as a regional reference point for stakeholder strategies or academic studies that encompass one or more countries in the region, the relationships between them, and relationships that exist between them and others. In this book, we treat the concept of the region in a similarly loose way. While respecting geographical definitions of where the region lies, we do not seek to be either comprehensive or all-encompassing in our treatment. Our focus extends across the region because we are interested in how and where environmental issues, screen media, and public art are implicated and how these practices in turn redefine the region. For this reason, however, our focus is also tempered toward the countries, sites, and cities that have been foci for the manufacture, consumption, exhibition of, and engagement with media and art. This has led us to focus on Australia, East Asia, South Asia, Southeast Asia, and

Japan, where flows and entanglements between the production and consumption of ICTs entangle artistic debates around the environment in particular ways, which are specific to those contexts but have implications across the region.

In the last fifty years or so, the Asia-Pacific region has undergone a series of changes that inform the way it is conceptualized today. In the 1970s, a widely held view cast the region as a bloc of "newly industrialized countries" (such as Taiwan and South Korea) oscillating around the region's first industrial nation, Japan (Dirlik 2007; Hjorth 2009). The constitution of the Asia-Pacific as a region and site of consciousness coincided with (and was indeed spurred by) the European quest for new resources from the 1970s onward (Dirlik 2007). At that time, the Australian prime minister Paul Keating actively constructed policies and strategies to strengthen Australia's relationship with Asia through the notion of the Asia-Pacific.

From the vantage point of Australia, the hyphen in "Asia-Pacific" represents two dominant geopolitical and ideological agendas. On the one hand, Asia represents the source of production of technologies, ideologies, and other forms of "soft power." Defined by Joseph Nye (1990), soft power is the concept that persuasion occurs through co-opting rather than coercion, attraction rather than force. The term has frequently been used by the United States to describe growing powers in the region, including China and South Korea.

The Pacific, by contrast, has often been imagined dually: on the one hand, as a site for the extraction or consumption of natural resources; on the other hand, as a destination for the export of ideas, values, and ideologies around development and governance (Tacchi et al. 2013). For example, countries such as Papua New Guinea have been the source of gold and phosphate mining over the last two centuries, and the beaches, coral reefs, and beauty of other countries such as Fiji and New Caledonia make them primary destinations for tourists. Simultaneously, flows of capital in the name of development circulate around the Pacific. Heather Horst (2013), in her ethnographies of mobile media and money in the Pacific, has explored how the Pacific has been imagined within the Asia-Pacific rubric as the site where money goes rather than where it is produced. For example, in 2011–2012, Australia sent more than $1.16 billion, or almost 25 percent of the country's total development assistance, to the Pacific.

In recent years, the Asia-Pacific has particularly been viewed as a site of rapid change and emergent forms of economic and political power. This has generated business and political interest in some countries in the region in particular. For instance, over the past decade, China's position has shifted from being regarded as a center for technological manufacturing to being one of the world's key sites of soft power. Indeed, some have denominated the twenty-first century as the "Asian century" (Arrighi 2009).

However, this denomination is also contested. As Arrighi highlights, thinking of the twenty-first century as the Asian century neglects the significant role of Asia historically. Yet beyond the academic debate, for neighboring countries in the region such as Australia, these new perceptions of Asia have generated concerns and questions about issues and opportunities that might emerge from engaging the region (see Beeson 2013).

The Asia-Pacific thus reconfigured has become a research site that poses new questions around old research themes. These include articulating and contesting local identities and transnational flows of people, media, goods, and capital (Arrighi 1994; Arrighi, Hamashita, and Selden 2003; Dirlik 2007; Wilson 2000). New Asian consumer cultures have emerged with the intensification of capital in the region and growing middle classes (Iwabuchi and Chua 2009). By the turn of the century, Leo Ching suggested that "Asia has become a market, and Asianness has become a commodity circulating globally through late capitalism" (2000, 257). In response to this trans-Asian cultural studies approach, scholars such as Kuan-Hsing Chen (2010) have advanced the idea of Asia as method—an idea that originated from a lecture that Takeuchi Yoshimi delivered in 1960. Chen uses the concept to think about Asia not as an object for or of other countries, especially the Euro-American West or imperialist Japan, but as a definition that comes from within Asia.

It is within this context that the study of screen ecologies in the Asia-Pacific needs to be understood. Indeed, to put it another way, contemporary screen media are part of the Asia-Pacific. The mobile phone has long operated as a symbol of emergent lifestyles (Robison and Goodman 1996) and likewise cannot be separated from the development of consumer cultures in the Asia-Pacific. In *The New Rich in Asia* (1996), Robison and Goodman identified the mobile phone as an index for increasingly common transnational consumption and new narratives of modernity in the region. More recently, with the emergence of increasingly affluent middle classes, albeit unevenly distributed across the region, mobile media have become part of the region's consumer culture (Hjorth 2009). Mobile media practices play out on both material and immaterial dimensions—from the customizing outside the phone to the personalizing of social media and photographs within the phone (see fig. 1.3). Simultaneously, accompanying the growth of this burgeoning consuming middle class are new forms of ethical consumption in which the ecological aspects of both production and consumption come to the forefront (Lewis and Potter 2011).

As a region, the Asia-Pacific thus represents a set of core concerns for the questions we discuss here. One of these concerns is how the demand for digital screen technologies drives resource mining, causing environmental problems. There is a deep irony

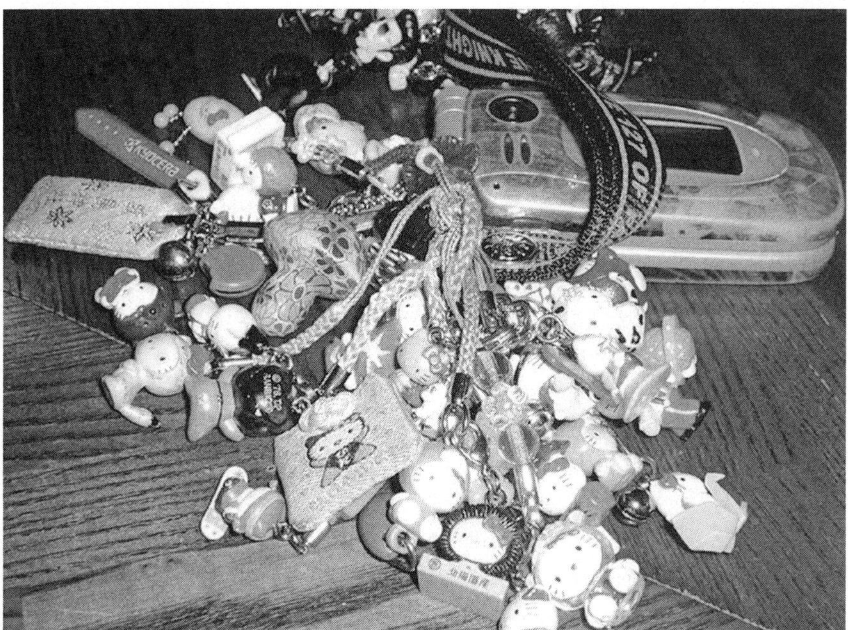

Figure 1.3
Larissa Hjorth, *Locating the Mobile*, 2004. Courtesy of the artist.

here, entangling art, technology, and the environment: technologies cause environmental damage, and yet, as we see in the case of artists, the same technologies can also be used to counter that environmental damage. Moreover, as we shall see, artists often deploy waste to comment on waste. A second concern is how global inequalities structure human labor and capital flows, with the result that much technology manufacturing is located in the region, along with the environmental pollution it creates. Third, newly established and emergent middle classes who benefit from the region's economic growth begin to increasingly use digital and social media, and this results in the growth of new consumer cultures.

As we outline in the following two sections, the art and media cultures that are being generated in the region offer us ways of thinking about and mapping the Asia-Pacific that both build on and depart from its conventional identity as a site for resources, manufacturing, and new consumer markets. Indeed, the new perspectives derived from viewing the region through its public art and digital screen cultures form the basis for the critical perspectives we build throughout the book. If we are to understand the implications of the emerging relationship between media and environmental sustainability "from within," then the present configuration of the Asia-Pacific region offers us an ideal example through which to explore this question. Dirlik has proposed that "developments in and around the Pacific have played a major part in the production of the phenomena that call for new paradigms" (2005, 165). Viewing the Asia-Pacific as a region that poses new analytical questions offers a way to advance a thesis for what Maxwell and Miller (2012) call "greening the media."

In the next sections, we set the scene with a series of examples of how the Asia-Pacific has emerged as a site for environmental engagement through public art and screen media. We present these examples as starting points through which readers might begin to gain a sense of the media and art contexts that form the subject matter of our discussions throughout the book.

The Asia-Pacific as a Site for Public Art

One of the ways in which the Asia-Pacific has become marked out that is of particular interest to our discussion is as an arts region. The Asia-Pacific has attracted significant attention through the rise of regionally focused art biennials and triennials such as the Asia Pacific Triennial of Contemporary Art (APT), Yokohama Triennale, Gwangju Biennale, Biennale of Sydney, and Shanghai Biennale, particularly since the 1990s. Held every two (biennial) or three (triennial) years, these curatorial exercises operate to remap the region through art in different ways. They have been accompanied by debates regarding the conflation of art and political economy and the role of art in globalizing national identity (Gardner and Green 2014).

An early example of this shift in art cartographies in the region was the Fukuoka Art Museum. Located in the west of Japan, near South Korea, Fukuoka was one of the first museums dedicated to collecting the art of the region. The founding of the museum and the exhibitions in the late 1970s showed forms of "local experimentation" that were a central part of the new intercultural networks developing among regional cultures. Following Gardner and Green, we can interpret these biennials as a counterforce of globalization and Western hegemony. Gardner and Green argue that the growth of biennials as a part of tourist routes saw them shift from local experimentation toward creating a form of *global legitimization*. They suggest that "the 'world art' model of cosmopolitanism developed a strange, vampiric afterlife, in which the conflict was temporarily resolved through the mutually parasitic, symbiotic importation of 'global' cultural knowledge by the singular, auteur, globe-trotting curator" (Gardner and Green 2014, 110).

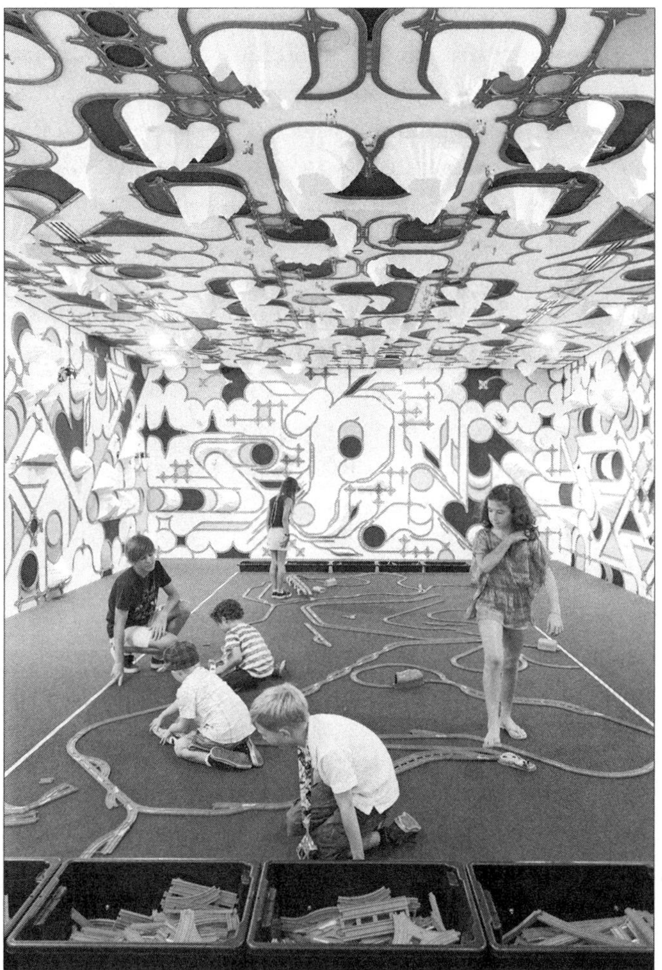

Figure 1.4
Paramodel, *Paramodel Joint Factory*, 2012. Image courtesy of Queensland Art Gallery, Gallery of Modern Art. Photograph by Mark Sherwood, QAGOMA.

While the biennial model has come under scrutiny and debate for replicating older models of colonial and Eurocentric structures, or for colluding with economic and political national aspirations, regional networks have offered significant opportunities for artists beyond national borders (Antionette and Turner 2014; Maravillas 2006). Such examples include Utopia@Asialink in Australia, Sa Sa Bassac in Cambodia, and Sàn Art in Vietnam. These new platforms emphasize art as knowledge production. They are critical hubs for social and political discourse, such as on environmental issues, and offer alternative structures and contexts away from mainstream museums for the public to engage with art. These hubs are key sites where domestic and international interest in contemporary art can intersect, and they are playing a critical role in reimagining the region by emphasizing collaborative and cross-institutional interactions. The hubs provide an alternative model to the templates offered in much of the rest of the world where residency, funding, and exhibition models are often weighed down by conventional gallery structures and institutions. They facilitate a more adaptable and flexible approach to art, encouraging social engagement and political participation, in the vein of what the art theorist Grant Kester (2004) calls "dialogical aesthetics." Kester's notion focuses on the interaction between dialogue and context rather than on an artwork (as object) in context. This approach—which directs art toward creating new knowledges through audience participation and dialogue—has been engaged by many art activism initiatives in the region. For example, the WAWA Project, a Japanese arts collective (discussed in chap. 8), overlays online media, art, and social workshops to strengthen emotional and community connections.

The BOAT PEOPLE Association (BPA) in Tokyo offers another example. Managed by two artist-architects, BPA works with local communities and public civic institutions to encourage reflection on the urban environmental context, in particular the social, physical, and economic ecologies of Tokyo's river system. Microcommunities are formed for each project, attracting local communities and art world publics to each event, who are often brought together through social media platforms. Such emerging spaces for contemporary art practice are responsive to their immediate contexts and the challenges and opportunities they afford.

For Zoe Butt, director of Sàn Art, the gallery provides a space in which to question Vietnam's highly regulated online environment, where, Butt notes, all public cultural activities, including visual art, performance, literature, and film, are subject to surveillance by the Vietnamese Cultural Ministry and Cultural Police. For Butt, the pivotal questions are how to build an engaged and active critical audience for art in a community suffering from self-censorship, how to use suitable promotional language, and how to strategically use websites and social media tools in ways that enable artistic organizations to evade the radar of local officials while being able to remotely contribute to international critical arts discourse (Butt 2014).

Both through the shifts and historical rewritings of art in the globalized context of biennials, and in the development of regional networks focused on hubs, we can see how the ways in which art is practiced and experienced in the region have become connected and revised into new configurations. These configurations begin to lend a lived identity to the region as a site for art and, as such, for art that is of concern to the region. This art engages with environmental concerns that are pertinent to the Asia-Pacific as well as having global relevance.

Screen Ecologies in the Asia-Pacific: Contexts and Snapshots

The ways in which the environment is represented, discussed, and experienced through media have changed dramatically over the past decade with the rise of social and mobile media. At the end of the twentieth century, as Anderson (1997) has discussed, "packaged" news media presented stories about the environment to the public. Such mass media representations of ecological issues in TV or print media provided little avenue for discursive debate and critique. However, since the 1990s, the rise of participatory mobile, social, and locative media has radically changed the ways in which ecological issues are discussed, presented, debated, and revised. Anderson's more recent work builds on Manuel Castells's (1997) efforts to situate environmental media in the context of a "network society." She suggests that mobile and participatory media are leading to a new immediacy with the rise of blogging, social networking, wikis, and podcasting. This brings news to us faster and through multiple sources (Anderson 2014, 4). Anderson further notes how these media are increasingly being used by environmental NGOs (6).

In the Asia-Pacific, increased capacities for dissemination, user-created content (UCC), and debate mean that networked screen media are now playing a pivotal role in how the environment is presented and how news and campaigns relating to it are disseminated and accessed. Networked and locative technologies, such as camera phones and smartphones, combine with microblogging and social media platforms (including Twitter, Facebook, and the Chinese Sina Weibo) to constitute new digital and online media ecologies that are multiple, contested, dynamic, and ever changing. Yet it is also important to remind ourselves how these media and networks are unevenly distributed across the region. For example, in some locations such as Japan, mobile Internet has been a part of the everyday since 2000. In South Korea, the introduction of strong broadband infrastructure in 2000 saw the country become a leader in online gaming production and consumption.

As Michael Arnold et al. (2014) note, the uneven implementation of the Australian broadband scheme, the National Broadband Network (NBN), can be understood as a "patchwork network" whereby spectrum politics play out in and through Australian homes. Australia is one of the few countries where users have to pay for the amount of information they download and where speeds are still slow. This has led users from other countries such as South Korea to conclude that the Internet in Australia is not *the* Internet (Hjorth 2007). Within the region there is also variation in the ways and extent to which the Internet is accessed. In Southeast Asia, mobile technologies like SMS have dominated, especially in places like the Philippines, which was lauded as the texting capital of the world because of cheap texting plans (Pertierra 2000). However, such developments need to be understood within the context of their national, infrastructural, political, and economic circumstances.

For instance, in the case of Indonesia, Citizen Lab notes how "despite its impressive growth and numerous small and large ISPs, Indonesia's ICT development has lagged behind many of its regional neighbours" (2013, n.p.). Malaysia's Internet penetration rate was close to 66 percent in 2012, while the percentage of individuals using the Internet in Singapore was 74 percent in the same year. In contrast, a "low level of bandwidth and computer penetration in Indonesia" (Citizen lab 2013) has limited the nation's process of increasing Internet penetration. Thus these differences also add to the ways in which, and reasons why, Internet platforms are likely to be engaged with in different countries, and how these will vary and change in multiple ways.

It is in times of political or environmental crisis, or through events such as those generated by public art (such as those discussed in this book), that we often observe the intensification of digital and screen media practices around public issues. Screen practices are always contextual, and as the examples we discuss make clear, each instance is framed by media ecologies specific to issue-based, political, and economic elements of national and regional contexts. Yet the following examples, each emerging from a different national context, demonstrate that screen media across the region are tied up with environmental issues and how they are disseminated, debated, and experienced.

For instance, the Fukushima nuclear disaster of March 2011 in Japan (known as 3/11) was accompanied by an intensity of social media activity, including news dissemination, debating platforms, lobbying, and public art. For some, this functioned more like a conference than a conversation (Hjorth and Kim 2011), as social media platforms such as Twitter, Mixi, Line, Kakao, and Facebook not only disseminated the unfolding events globally but also served as media platforms for global debates about nuclear power. Counterpublics and local lobby groups used mobile social media to participate in these discussions and were therefore able to gain a global stage in the face of governmental hegemony. Artists also engaged mobile social media to connect people and issues related to the disaster. For example, art media projects like WAWA at 3331 Arts Chiyoda (Tokyo) evoked dialogical aesthetics in an effort to use art and new media to help connect and fund those affected by the 3/11 aftermath.

In China, a country whose politics and media policies have long been the focus of critical global debates about democracy, artists such as Ai Weiwei and Cao Fei have used social and online media to explore issues relating to politics, the environment, and creativity. In a context where media use and access are tightly regulated, social and digital media are navigated differently, bringing together art, media, and activism through configurations that connect the local and global in novel ways. For example, Ai Weiwei is famous for using Western media like Twitter and Facebook, which are banned in China, to generate debate and discussion about China. Cao Fei has used the globally accessible online digital platform Second Life as a performative space. In the

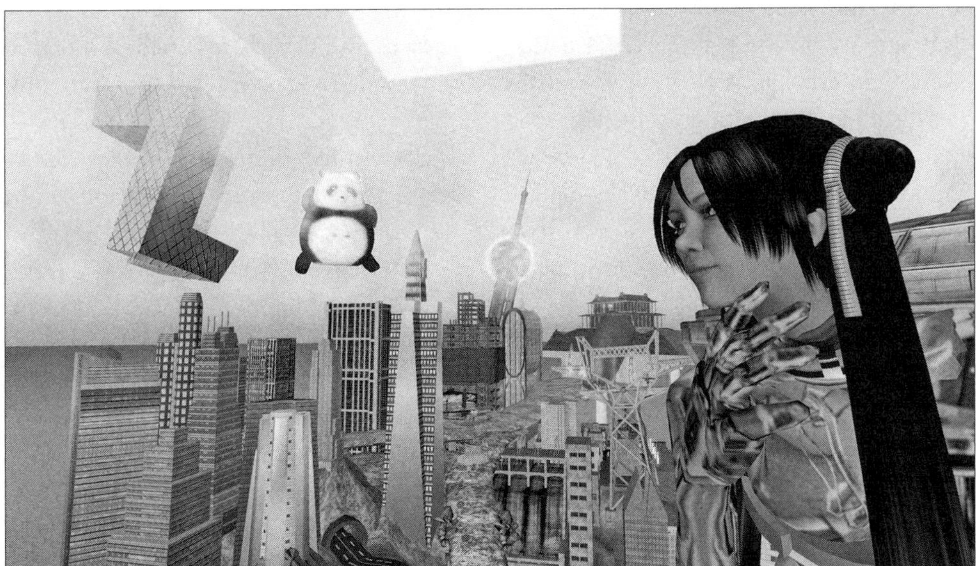

Figure 1.5
Cao Fei (Second Life avatar: China Tracy), *China Tracy and RMB City*, 2007–2011. Image still from RMB City. Courtesy of the artist and Vitamin Creative Space.

online world of Second Life, Cao Fei has created an avatar that traverses transnational popular culture images to invert South–North global flows and to reflect on debates about the commodification and mediatization of the environment. Cao Fei's exploration of the role of consumption as cultural practice is much more ambivalent than the overtly political work of Ai Weiwei.

In Australia, another rather different context, art networks have played key roles in local, regional, and global conversations concerning art, screen-based and online media, and the environment. As home to the region's first green political party (the United Tasmania Group, precursor to the Tasmanian Greens, was founded to save Lake Pedder in 1970) and one of the first countries to found a branch of the globally influential Greenpeace (in 1971), Australia has a rich history of environmental activism. As Linda Williams observes, the representation of the environment through the photographic lens has played an important role in activism. Pointing to the use of documentary photography in the successful protest from 1976 to 1983 against the Franklin Dam, Williams argues that Peter Dombrovskis's *Morning Mist, Rock Island Bend, Franklin River* (1979) is one of the most celebrated landscape photographs in Australian history. As she argues, it "was a persuasive public environmental image that made a substantial contribution to saving the Franklin River, and which left its indelible mark on the political campaign that brought about a change of government" (Williams 2013,18). Photography plays a key role in visualizing the tacit, and this is amplified in a period of camera phone apps whereby romanticized, analogue-looking images of nature proliferate alongside the billions of selfies.

In 2008, RMIT University Gallery in Melbourne hosted the *Heat: Art and Climate Change* exhibition, curated by Williams. This was the first survey exhibition of art responding to climate change in Australia (the year before the important *Earth: Art of a Changing World* at the London Royal Academy of Arts in 2009). In 2012 it was followed by *2112: Imagining the Future*, also curated by Williams, which focused on how artists envisaged the environment in a century's time. Other examples, such as Greenpeace Australia's online game *Ocean Apocalypse* (2012), ad-busting videos like *Reject John West* (2012), or other sites such as *Tuvalu Climate Change SOS* (2008), are indicative of how Australia's artists have played a key role in marrying environmental politics, new media, and art in the region. Many Australian artists exhibit works online, drawing attention to local ecological issues, and the emerging Australian artist Josh Wodak has adapted his new media art practice to highlight the threat that rising sea levels pose to Pacific islands such as Tuvalu.

In 2012 the Australian Network for Art and Technology (ANAT) collaborated with the Melbourne-based Carbon Arts to run a competition called *Echology: Making Sense of Data*. The competition signaled the growing significance of public art and real-time screen media to engage with issues of sustainability. From Natalie Jeremijenko's *Mussel Choir* (a biological artwork deploying a mussel colony's movements as a way to understand water quality improvements in Victoria Harbour) to D. V. Rogers's *Terra Sensing Tower*, the potential for art and new media to connect and intervene in environmental issues was unmistakable. Deploying "sensing new media"—that is, the use of different senses, not just the visual—the projects gave audiences a chance to reimagine the environment as a dynamic space. In another project, titled *Curating Cities*, created through a collaboration between Carbon Arts, the University of New South Wales's National Institute of Experimental Arts, the City of Sydney, and Object Gallery, researchers examined "how the arts can generate environmentally beneficial behaviour change and influence the development of green infrastructure in urban environments" (Carbon Arts 2014) through online and offline interventions in the urban context.

Finally, in Taiwan, the impact of rapid technological progress has caused many artists and activists to consider the role of art as a vehicle for ecological debate. One example is the Cheng Long Wetlands Environmental Art Project, now in its third year. Over this time, social and mobile media have increasingly played a role not only in disseminating information and establishing visibility but also in stimulating creativity and debate. The Cheng Long project deployed social media such as Facebook, not just for dissemination but also to generate a type of parallel "dialogic aesthetics" in and around the project's ideas. In 2012 the Taiwan-based curator Jane Ingram Allen organized a touring exhibition of Taiwanese environment art titled *Going Green*. For Allen, the emergence of environment art has accelerated dramatically in Taiwan as the country tries to reconcile its rapid rise into twenty-first-century modernity and the impact of technological progress on the landscape. With Taiwan epitomizing the newly emergent middle class of the region, the exhibition's focus on intersections between ecology, art, and screen culture represents new kinds of knowledge, ways of knowing, and political activism.

As these examples from four different Asia-Pacific countries show, there is a growing relationship between the recognition of environmental issues and how screen media and public art are increasingly engaged, often in tandem, in publicizing, critiquing, and confronting these issues. While Japan, China, Australia, and Taiwan have different political systems and histories, as well as varied levels of Internet access and performance, they share in common the presence of affluent, technologically savvy, and educated middle classes who participate in active public art cultures. In many ways, they epitomize the conflicts that form the core of our investigation, bringing together an ironic mix of consumer culture and technological development with growing sectors of educated, critical, environmentally aware, and engaged publics.

Asia-Pacific Screen Cultures and Public Art: Concepts and Issues

As a region, the Asia-Pacific represents a series of core concerns for this book. It brings to the fore the relationships between resource mining for materials required to produce digital screen technologies; the global inequalities related to flows of capital and human labor needed to manufacture such technologies; the environmental pollution resulting from manufacturing; and the ways in which new middle classes with their concomitant consumer cultures create new demand for, and forms of waste associated with, digital and social media. The region is undergoing rapid urbanization, has the globe's largest percentage of megacities, and is the site of much ICT manufacturing and consumption. There is increasing (if not universal) recognition that the region's recent unpredictable and devastating weather patterns are related to global environmental change, which in turn has generated the thriving screen media and public arts cultures that we began to map out in the earlier sections of this introduction. This combination of factors offers us an ideal example through which to ask a set of key questions about how the relationships between environment, screen media, and art are emerging, and how they might be harnessed for a sustainable future.

By researching these relationships throughout the Asia-Pacific, *Screen Ecologies* both complicates and advances existing debates and assumptions about how environmental movements and critique work within a network society (Anderson 2014) and about the greening of the media (Maxwell and Miller 2012). In the next chapter, we outline a framework for thinking about these different elements and the relationships between them. First we set out the structure of the remainder of the book.

An Outline of the Book

Each chapter of the book takes a different angle on the intersections between art, media, and environment. In chapter 1, we locate our motivations for the book and how overlays of art, media, and environment allow us to reimagine the region. In particular, we highlight the role of artists in exploring alternative ways in which to visualize, practice, and represent the environment. In chapter 2, we set out an approach for examining the relationships between art, media, and climate change, drawing on anthropology, media studies, and cultural studies.

In chapter 3, we connect artistic explorations and representations of the environment with stories of media innovation and climate change in the region. To do so, the chapter contextualizes global media debates in relation to some of the emergent environmentalism issues in the region. Then we turn to examples of artists working through these issues, noting the undulating connections and disconnections between media, art, and the environment.

In chapter 4, we consider how mobile media require us to rethink the types of media and capital flows that take place in the Asia-Pacific and what these movements say about cultural geographies. As noted earlier, the rise of mobile media has become synonymous with screen culture in the region. Mobile media have also played a key role in amplifying, as well as disrupting, media capital flows. In this chapter, we address the tensions between amateur and professional art in relation to media technologies and how they affect perceptions of the environment. We finish by reflecting on the ubiquitous role of camera phones in these contemporary visualization cultures.

Chapter 5 shifts the discussion to art and its relationship to the emergence of the "platform" as a concept and place through which artists engage across the region. Focusing particularly on the role of public engagement, this chapter looks at the rise of transregional spaces in contemporary art (such as biennials) and how they participate in making regional ecologies of place and forms of Asia-Pacific cosmopolitanism. Tracing the rise of regional biennials and triennials, transregional projects, and new forms of gallery and arts organizations, we reflect on how these have brought new critical attention to the region's art practices and politics, and how they in turn are implicated in global mediated and economic flows.

Chapter 6 examines how contemporary cultural and local debates have engaged in effective environmental critique. It focuses on the strengths and limitations of art and social media as an effective method for rethinking climate change. In particular, this chapter considers how regional art responds to the theme of water, such as in the Australian-based *Spatial Dialogues* public art project, Touchmedia's Eco-Art project, Chen Qiulin's *River, River* project in China, and Japanese projects such as the BOAT PEOPLE Association (BPA) and the Green Island Project.

In chapter 7, we outline new models for collaborative cross-cultural art projects in the Asia-Pacific with a particular emphasis on how networked and social media can be used to reimagine artistic subjectivity and engagement and provide new paradigms for understanding collaboration in a global context. The use of multiple platforms for engagement across a diverse range of geocultural spaces is a particular focus in analyzing the complex process of exchange and dialogue in collaboration. Our case studies include *The Library* by Soundpocket and *Stereopublic.*

In chapter 8, we consider how, while social and mobile media exemplify types of publics in which intimacy is mobilized and commodified (Illouz 2007), they are also moving from private to increasingly public realms. In chapter 9, which concludes the book, we summarize how the intersection between art, media, and the environment in the region allows for new ways to reimagine the Asia-Pacific through changing depictions of the environment. To end, we reflect on the relationship between arts practice and critical inquiry through a series of proposals concerning how new media arts might usefully be harnessed to intervene in debates about climate change in the Asia-Pacific.

2 Media, Art, Climate, Publics

To approach a problem as urgent as climate change, and to understand the past, present, and potential future place of digital media and art in relation to it, we need a methodology that can accommodate the trajectories of each of these themes *and* also bring them together. In this chapter, we propose an approach to the world that situates art, media, and climate change—and ourselves as humans—as part of the environments that we inhabit. This approach views what is happening around us from the *inside* and explores the relationships between the things and processes that constitute the world.

To conceive of these rather different things and processes—media, art, climate change, and humans—as part of the same configuration involves focusing on the ways that they affect and shape each other. Yet conventionally, each of these aspects has tended to be theorized differently by the various disciplines that take them as their object of inquiry. Therefore the theoretical problem of understanding media, art, climate change, and human activity in the world is an interdisciplinary task that needs to account for the work of disciplines including media studies, cultural studies, art, and anthropology, as well as to navigate their respective sensitivities and priorities.

To achieve this goal, we introduce a series of concepts that set the stage for our analysis in the following chapters: environment; representation and process; artifact and event; place and platforms; movement and mobility; and publics and intimacy. In each case, we advocate an approach to the research problem and the world or environment that we inhabit from the *inside*. In the next section, we outline how these concepts enable our analysis, before explaining each in more detail.

Conceptualizing Region, Climate Change, Media, Art, and Publics

Regions are constructed by academics, politicians, and cartographers, to name but a few interlocutors. They are also constituted and represented through art and media practices and content. Their significance and shapes shift over time as units of political administration and relevance, economic consolidation and unity, and analytical interest to researchers. This very observation implies that regions are difficult to catch and to hold on to. While on the one hand this shifting makes the study of regional ecologies of climate change, screen media, and art a slippery task, it also reminds us that this book is not about the study of fixed or static units but about how certain configurations of things, processes, persons, technologies, and representations come together as part of a processual world.

Studying the Asia-Pacific region provides a way to understand how processes of climate change, screen media technology and its use, and art practice are converging or relational to one another, thereby presenting an alternative way of approaching this issue. It means being open to findings, contexts, and environments that do not correspond with the expectations framed by theoretical approaches based on the study of Western modernities and their technological and climate histories. Investigating these questions through the Asia-Pacific does not simply add to our knowledge about media, art, and environment in the world but addresses the question from an alternative perspective. Some of the same questions have begun to be addressed in existing literature. For example, Maxwell and Miller (2012) have taken a political economy approach to reveal the ironies, inequalities, and interdependencies of the relationships between media and environmental issues. Anderson (2014) has examined the changing relationship between media and environment in the sociological context of a "network society." Their inquiries are equally pertinent for the Asia-Pacific.

However, we argue that a more interdisciplinary approach is needed to understand this relationship fully, for three reasons. First, we are concerned with how digital media and art are used to represent, communicate with, engage with, and contest climate change and environmental issues. Therefore we need to consider how screen media and art can be made to stand for and critique the specificity of local everyday, infrastructural, political, participatory, and economic processes. This focus on representations, the content of media images and art, and the ways in which they might circulate and generate meanings in the world has largely been the subject matter of Western scholars of visual, cultural, and media studies (e.g., Hall 1997; Wells 2003; Mitchell 2005). More recently, scholars such as Mirzoeff have turned the techniques of this analytical approach toward climate change images, as discussed in the next chapter.

Second, screen media and art are *part of* the environment, part of a *changing* environment, inextricably embedded in global processes and flows. Their manufacture is responsible for the very carbon emissions argued to be leading to climate change. This becomes apparent not only as it is represented but also in the ways it is experienced.

This acknowledgment turns our attention to the nonrepresentational (Thrift 2008) or "more than representational" (Lorimer 2005) elements of change and of the relationship between media, art, and environment, where media and art, and the processes through which they are made, are *part of*, and contribute to, the making of a changing environment (see also Ingold 2010). Such approaches have developed in the fields of anthropology and human geography, where the focus has shifted away from the question of what is *represented* to that of how we experience and perceive the worlds we are part of.

Third, screen media and art are also being used to stimulate particular types of social, material, and political critique and initiatives in relation to climate change, and to involve publics in these processes. To understand the implications of this, we need to focus on how people personally experience media, art, and climate change. Recent work by scholars in cultural studies has pointed to the emergence of new forms of intimacy, based on more personal and emotional sentiments and engendered through mobile screen media. We argue that these sentiments are critical for our understandings of how the dissemination of media art can be an effective way of engaging publics with issues relevant to climate change.

As we move toward the final chapters of the book, which focus on how participation and publics are generated around media art, the concepts of intimacy and publics become increasingly central in showing how mobile and intimate publics are emerging and becoming active in relation to climate change. These concepts offer us a way to understand both the experiential and representational aspects through which climate change, environmental issues, and their critiques are being portrayed through screen media and public art, thereby becoming part of contemporary public worlds and everyday lives.

In the remainder of the chapter, we outline a way of thinking about the relationship between art, media, climate change, and their publics that contests conventional tendencies to see art and media as simply representations of the environment or of climate change.

Environment: Climate and Weather

The concept of *environment* has been used to mean many things in existing research, depending on discipline or political orientation. The idea of "the environment" is invoked in materials discussed in this book as something that exists externally to humans, media, and art in the form of "environmental challenges," the "environmental costs of dams," environmental "risk," "decay," and resistance (chap. 3); as something that is impacted by the media industries; as a mobile game that "reflects upon the environment" (chap. 4); and as "environmental politics" (chap. 5). Presented as such, "the environment" becomes a metaphor for something that is "other than" us and that we have a responsibility to act on, act for, and represent. It is presented as an entity that is separate from us, one that is being damaged or protected by politics, ruined by the disastrous effects of carbon emissions or pollution, and represented and campaigned for through media and art.

Throughout the book, while we inevitably do look *at* the environment and consider what is being *done to* or *done for* it, we also critique this perspective. The construction of "the environment" as a representational category is just one way of constructing a problem and one mode of confronting it. That is, it offers us a set of certainties about there being an environment "out there" that we can try to save. This approach *separates* the environment from people's actions and representations of it and offers us a way to consider how human designs for resilience against, and mitigation of, climate change might be thought out. It also offers a way to consider the politics of climate change, the representational categories through which it is debated, and how art and media are implicated in these. It means that while it is understood as part of a representational world, the environment is something that is conceived as *representable*—something that art and media can comment on and make things "stand for," something that the carbon emissions of mobile screen media manufacturing can damage, and something that, as academics, we might draw attention to. However, this concept of environment is increasingly being interrogated and critiqued.

For instance, in the field of ecocritical theory, references to the term "the environment" have gradually ceded to privileging the dynamics of processual ecologies. The ecocritical theorist Tim Morton (2013) has contended that while terms like "the environment," "the biosphere," and "global climate change" are necessary parts of communication, they are nonetheless rapidly becoming what he calls "hyperobjects." Morton sees such hyperobjects as being massive, nonlocal, and operating on non-human timescales. This makes them concepts that are simply too vast, complex, or indefinite for us to grasp, so they prompt a crisis in conventional approaches to understanding. Yet there is another way of understanding the environment that, we propose, is useful as an analytical and organizing concept. As the anthropologist Tim Ingold puts it:

We are these days increasingly bombarded with information about what is known as "the environment" … that we are, I think, inclined to forget that the environment is, in the first place, a world we live in, and not a world we look at. We *inhabit* our environment: we are part of it; and through this practice of habitation it becomes part of us too. (2010, 95)

Following this approach, we must see the environment not as separate from us but as inextricable from our everyday activities. The environment is not an abstract entity that we can do things to from a distance or something that we can "save" from afar. Moreover, it is thought of as being so "vast" that it is difficult to "grasp." The environment, therefore, is an ecology that we are part of and have a mutual relationship with. It changes as we change, and we both shape and are shaped by it. Such a processual, nonrepresentational understanding of environment is connected to a different understanding of human agency, resilience, and the relationality of things (Pink and Lewis 2014) than the focus on representational categories that we outlined earlier. It invites us to treat both human agency (see Ingold 2013) and interventions in the world as emerging from a processual world that they are part of, rather than being planned out or preconceived to *affect* a world that is separate from them. This theory also allows us to understand the relationship between climate change, art, and media in ways that appreciate them as *part of the same environment*, whereby environment is not separate from the people and things that constitute it. Here we cannot *act on* the environment as if it was an entity that exists separately from us. Rather, we are continuously involved in its making through everyday practical activity as artists, activists, and audiences.

An important conceptual step, therefore, is to consider how the notion of environment can stand for something that is constituted through the representational practices of humans, *as well as* for something that we, the weather, art, and media are all part of. A helpful distinction to use when thinking this through is Ingold's remark that "whereas the globe is measured and recorded, the environment is experienced. One has climate, the other has weather. One has its atmosphere, the other includes the sky" (2010, 96). Ingold's point is that "the environment" is the measurable and representational context in which climate is constructed and is said to change. Yet the experiential world that we inhabit is one where the weather is the means through which the category of climate change is made "real." When we experience weather conditions that are increasingly extreme, unusual, and erratic, people begin to correlate *the experience of weather* with the science of climate change. Thus the experiential and representational worlds meet to create convincing arguments that we are witnessing and *experiencing* the consequences of climate change, some of which we have already invoked in chapter 1 as part of our own argument that climate change is happening and needs to be addressed.

Screen Ecologies is written at the intersection where these different perspectives meet. This brings us to the complication that the book is therefore written at the intersection of two ontologies—that of the representational measured environment, where climate is debated, and that of what Ingold (2010) calls the "weather world," where we experience, feel, and engage in practical activity around media, art, and weather. Herein lie the participatory processes through which people engage with public art, play online games, and run through the rain. Of course, these worlds cannot completely be distinguished from each other, since the work of climate science is as much a crafted human activity as making art and making media.

We engage the notion of place as an organizing concept through which to think about how to bring together the experiential and representational within one framework through which we can then relate concepts of space, scale, time, movement, and perception. We are interested not in the idea of place as locality but in how a theory of place can help us to situate localities, technologies, weather, representations of climate or environment, media, and art (and other things) as relational to each other. Therefore the theories of place that inform our understanding in this book involve a focus on relationality between the different "things" that constitute place and the implications of these relations. The geographer Doreen Massey's notion of place as event helps us to think about this when she writes:

This is the event of place in part in the simple sense of the coming together of the previously unrelated, a constellation of processes rather than a thing. This is place as open and as internally multiple. Not capturable as a slice through time in the sense of an essential section. Not intrinsically coherent. (2005, 141)

We suggest that a theory of place might be used to imagine how things and processes of different qualities, such as climate change, screen media, and art, become relational to one another and what the effects of this can be. Invoking Maxwell and Miller's (2012) work again, we argue that the politics of media and environment are fundamental, and a theory of place enables us to situate these politics to understand how they play out both in specific localities and in relation to other global and regional flows. By both seeing place as a politically imbued event (following Massey 2005) and accounting for the notion of "place beyond place," we can better understand the implications of what happens in one place for others (Massey 2007). Such a theory of place moreover engenders a sense of responsibility in terms of imagining the consequences of actions in one place on another (as Massey indeed suggests).

This approach can help us to think through the question of how climate change, screen media, art, and the processes through which they are produced are entangled in complex meshworks (Ingold 2008) and are relational to one another in ways that are technological, infrastructural, political, representational, participatory, and experiential.

Representation and Process: Digital Screen Media

As we have noted, the concept of environment is complicated. We need to deal with "the environment" as a representational category, as well as with the environment as something we are part of and constitutive of. Screen media present us with the same complication, since to understand how screen media work in the world, we need to acknowledge that they have a purpose and role through their content and communication capacities as *representation*. Yet screen media play other significant roles in making the places that we are also part of: they are part of a processual reality whereby the carbon emissions from their production and use are leading to climate change, and they are part of the way we experience and navigate the world through their presence and the copresence of others through them (Pink and Hjorth 2012; Pink and Leder Mackley 2013).

Cultural and media studies approaches have conventionally focused on media's qualities as representation and their possibilities for communication and dissemination. Some media scholars, including Hjorth (2009), have tackled this issue by engaging notions of the social construction of technology (SCOT) (as developed in Bijker, Hughes, and Pinch 1987). A SCOT approach emphasizes the agency of human and social context in informing the use and uptake of technology, thus arguing that technologies must be understood as embedded within a social context. This also enables us to focus on how the meaning of media technologies is constituted through their use, in ways that are akin to domestication theories (e.g., Silverstone and Hirsch 1992). In *Screen Ecologies* we engage with this SCOT perspective but also build on these ideas to further situate media as also part of, and capable of, shaping a wider environment of things and processes. This, as we argue throughout the book, includes the proposal that media technologies and their social constructions and uses are equally shaped by, and constitutive of, processes of climate change, environmental disaster, *and* potentially also their mitigation.

Over the last decade, media scholars have become increasingly interested in theories of place. Like the framework outlined in the previous section, these approaches have tended toward both a phenomenology and a politics of place. For instance, more than a decade ago, Nick Couldry and Anna McCarthy called for a focus on both the materiality of media in that it is "composed of objects (receivers, screens, cables, servers, transmitters), embedded in particular geographical power structures ... and reflective of particular economic sectors in capitalism," and its virtuality in terms of what they call the "anti-concrete sense of spatiality" (2004, 3). These create what Couldry and McCarthy refer to as "complex entanglements of scale" that are "engendered by mediated forms of social coordination" (7). Their emphasis on the notion of entanglement resonated well with the work of geographers and anthropologists and invited what has become an increasing focus on place in media phenomenology as it has developed in both mobile media studies and non-media-centric media studies. Couldry and McCarthy's proposal revealed the geographic, economic, and capitalist frames that impinge on the roles that media play in the world, and remind us to attend to the infrastructures that enable media content, communication, presence, and participation; the affordances of these infrastructures; and the ways they might be shaped by, or resilient to, demand.

Scholarship in human geography has acknowledged the technological and infrastructural elements of place (e.g., Amin 2008; Thrift 2008) and highlights how the experiences of digital media that form part of everyday life are contingent on wider infrastructural systems that are often hidden from everyday view. These include cables and currents that flow underground (Amin 2008) and the mathematical (e.g., Thrift 2008) and algorithmic (e.g., Uricchio 2011) architectures that are inseparable from the actual forms of practical activity that we engage in with digital media. In the case of the Asia-Pacific, this also means accounting for the nature of national digital broadband infrastructures, bandwidth, and the different ways in which people normally access the Internet. Therefore the technologies and digital platforms through which digital participation can take place become part of the ways in which the politics of place are lived in everyday contexts. To understand this idea, we take an approach that attends both to the history of media in the region and to its varied uptake.

On the one hand, this approach entails mixing media archaeology (Huhtamo and Parikka 2011; Parikka 2012) with remediated methods (Bolter and Grusin 1999). Media archaeology is a particularly interesting approach in the context of this book because it is both a theoretical and an artistic methodology that examines earlier technologies to understand new media. By excavating older technologies and practices, media archaeology frames media in terms of continuities, rather than discontinuities, as the "new" in "new media" would lead us to believe. So, too, remediation methods take a dynamic and cyclic approach to technology whereby new and old media have a dialectic relationship rather than a simplistic causal narrative.

On the other hand, we account for the different narratives and configurations to be found across the region. As we noted in chapter 1 through a comparison of the Japanese and Indonesian contexts across the Asia-Pacific, the infrastructures, technologies, and modes of access available are highly varied throughout the region. Indeed, each context in which digital media use is developing is created in different ways, and we need to be attentive to this.

Theories of place have, in varying ways, also filtered into studies of mobile communication and the Internet (e.g., Hjorth 2005; Ito 2003; Morley 2005; Pink 2012; Wilken and Goggin 2012). For example, Mizuko Ito (2003) argued in one of the first ethnographic studies of mobile media use in Tokyo that the importance of place in media practices is magnified. Indeed, arguments about the contestation of place have become central in mobile media studies (Wilken 2005; Wilken and Goggin 2012). Anne Beaulieu (2010) suggests that ethnographies of online media have led to a rethinking of place by fostering a shift from the significance of colocation to that of copresence. Along these lines, Rowan Wilken (2005) also argues that networked mobility and mobile phone use are dramatically altering theories and understandings of place.

For the purposes of this book, we seek to go beyond both of these existing fields of study. We are concerned with understanding how place is constituted and the role of digital media in its constitution. However, our focus and starting point are not digital screen media themselves but their place in an intersection between media, art, and climate change. This point coincides with Couldry's complaint that one of the problems with media studies is that scholars in the field often tend to assume that media "are the most important thing in people's lives" (2012, x). Couldry redirects us to an approach that is "grounded in the analysis of everyday action and habit" (x). Likewise, Shaun Moores identifies with a "non-media-centric approach" (2012, 11) and "attempts to understand everyday media uses by considering them alongside many other social practices today, rather than as isolated activities" (x). Along with these theorists, we are seeking to understand media *alongside*. Yet, adding to their points that we need to see media uses as happening alongside and in relation to other "action and habit" (Couldry 2012) or "social practices" (Moores 2012), we propose an approach to media that is alongside other *processes*. It is not simply a question of situating what people *do* with media in relation to other actions, but a question of understanding media use as a process. Processes of media use are therefore relational to processes of climate change, art practice, and public engagement with art, and to the technological and infrastructural processes and flows that are producing climate change, on the one hand, and more digital screen media technologies, on the other.

Place is particularly interesting as an organizing concept for such a non-media-centric approach because, as understood here, it does not privilege digital media as an object of study. Taking place as an abstract model through which to conceptualize precisely that which is *not* abstract, it invites us to see digital media as *part of* a wider environment, interwoven and interdependent with other things, processes, and materialities. It invites a study of digital media in terms of (and as inseparable from) their relationality with other constituents of place, which might have different organic, physical, material, and sensory qualities and affordances. By taking a non-media-centric approach to the question of how media are part of the entanglements of place, we can go beyond the notion that digital media *do things to* our identities, relationships, and other elements of our lives.

This approach, moreover, does away with the idea that digital media can be understood as separate from, and as having the potential to influence, ways of being, such as having relationships with other people or comprehending who we are. Instead it sees digital media as embedded in the experience and making of the worlds that we are part of and the selves that we live out. Conceptualized in this way, digital media do not "do" things to us, to our identities, practices, or environments; rather, they become interwoven with our emplaced being and doing.

Artifact and Event: Art and Media

In the previous section, we emphasized how, because media are part of the environments we inhabit, we cannot actually separate screen media and climate change. They are emergent from the same configuration of things and processes. A parallel argument can be made in relation to art: in the contemporary context discussed here, the relationship between art, screen media, and climate change cannot be dissolved. It is not so much that one can act to change the other but that each is mutually implicated in making the environment that they and we are part of.

In this book, we unpack this relationship through the concept of the platform. Platforms are part of both the infrastructure and the conceptual frameworks of digital media and arts, although their material and experiential components are manifested differently in each domain. The concept of the platform helps us to address the relationships between the experience and representation of art and media, since both digital and art world platforms can be interpreted as *sites of representation* and, at the same time, part of the environments we actually inhabit *with* art and media.

Digital platforms are part of smartphone app ecologies and ICT use, as discussed in chapter 4. Social media and gaming platforms form part of the digital materiality of everyday life, given that the participants in the research projects we discuss both traverse and make digital representational content and experience the world through the entanglement of the electronic and material infrastructures in which they live. One way to understand how people engage with apps, platforms, and mobile technologies is through the related concepts of movement and place. If we recognize the processuality of the worlds we live in and are part of, it becomes impossible to see artworks or media screens as static objects.

Instead, we understand the technologies, artifacts, and persons who engage with them as being mobile and as part of a world in movement. In chapter 4, we develop this idea in relation to camera phone photography taken "on the move" and to mobile gaming through art. We introduce the concept of "emplaced visuality" to examine the embodied and affective ways that people engage with camera phone photography and participate in mobile gaming. As we show, they do so by participating in an environment where the physical-material and technological weather world and the digital-intangible and algorithmic world are entangled as part of the same environment. Apps and platforms are part of this environment and are fundamental to both the representational and infrastructural realities that we inhabit. They are technologies that we observe things *in* while at the same time being technologies that we move through and around.

The platform is also a key concept in the art world, as we show in chapter 5. We view platforms critically, following Tarleton Gillespie's point that their discourses might "matter as much for what they hide as for what they reveal" (2010, 359). We consider platforms as emerging as places in which the digital and material are entangled, and we account for how platforms are made through intensities of social engagement as much as through their occupation of physical sites. Platforms become central to thinking about the clustering of experiential and representational forms, as part of places that are made up of things, persons, and processes of different qualities and affordances, and as open, continually changing aspects of the environments we inhabit. The place-platform pairing therefore offers us a concept through which we can recognize the "eventness" of art and of media as they become part of everyday Asian screen ecologies. The platform offers us a prism through which to view the ways in which climate change becomes representation, discourse, and experience, as it is mediated, contested, and experienced, and how it indeed becomes an increasing part of the ways everyday life is lived.

Mobility and Movement

Mobility manifests itself in contemporary cultures and societies worldwide in multiple ways that are part of how the Asia-Pacific can be understood in relation to the rest of the world. We connect four key foci that are brought to the fore by the concept of mobility: flows of ideologies and critiques, the movement of people, the movement of things, and the notion of people always *in* movement. Bringing these ideas together, we emphasize the notion of movement as being core to our understanding of the world. Movement is a fundamental quality of the contexts that we research and discuss in this book. It is a quality that makes the world and our journeys through it incremental and cumulative journeys, forever collecting new experiences and creating new dynamics rather than ever backtracking or repeating what has gone before.

Chapter 6 focuses on the ways in which ecocritical art is participating in the Asia-Pacific region. Here we see how the movement of artworks, as well as ideologies and debates, from the Asia-Pacific plays a key role in defining how climate change becomes part of cities in the region. Indeed, this point is manifested in the field of ecocritical scholarship as we witness the emergence of a specifically Asian strand in this work. It is, moreover, one of the ways in which the ecocritical art of the Asia-Pacific and elsewhere can be seen as part of a wider and shared ecology of ecocritical works that become referential to one another as well as to the local, regional, and national issues that they respond to and direct in new ways. Thus these movements and appropriations of art, debates, and critiques are part of wider ecologies of the movement of ideas, the artifacts associated with them, the movement of people, and digital flows.

We emphasize the need to understand the emergence of new and critical forms of climate-aware public art and screen-based representations within their relationship to the wider ecologies of movement within which screen media are implicated. The movement of people, as migrants of all kinds, is an important element of these ecologies of movement. Human movement includes that of refugees, asylum seekers, temporary workers, businesspeople, tourists, travelers, and many more. They make traces and trails across, through, and within the region, often doing so in ways that leave a digital as well as human trace. This context is, as a good number of scholars and researchers have documented, one where digital screen media are inextricable from the paths, communications, and feelings of comfort that people who move between nations and continents experience while in movement.

Yet it is not only people who are mobile but also screen media that can become implicated in the global movement of people and things in ways that are surprising, such as a movement that intersects with old media and everyday materialities as well as connecting people and things in novel ways. For example, in 2014 a BBC News article described how social media accounts were used to trace a Cameroonian man who, when he was a prisoner in China, hid a plea for help and a photograph of himself in a bag made in a prison factory, which was found by an Australian woman in New York (BBC Editorial 2014). Seeing the Asia-Pacific as a producing and consuming region offers us a particular lens through which to consider how the movement of digital media, technologies, their components, knowledge, capital, and the workforces related to them are bound up in questions of climate, environment, power, and ethics.

Media phenomenologists are increasingly calling for us to attend to the experience of media (e.g., Couldry 2012; Moores 2012; Pink and Leder Mackley 2013), including mobile media (e.g., Farman 2012, 40). The work of Ingold is helpful here because his "overriding aim is to understand how people perceive the world around them, and how and why these perceptions differ" (Ingold 2011, 323). This approach is highly relevant for thinking about the sensoriality of a world in which digital media are increasingly ubiquitous (see also Moores 2012), and it offers us a way of grasping how people move around in environments that are at once digital and material, and of which screen ecologies are a part. Movement is therefore not simply the way we get from place to place; it is also fundamental to how we learn about and engage with our environment, including the digital and screen media and art forms that are part of it. We therefore need to understand mobile publics as also being made up of sensing, mobile people. These points are particularly relevant to the public art discussed in chapter 6, for example, enabling us to comprehend urban air pollution as both experiential for mobile subjects moving through cities and representational in the form of ecocritical art.

Digital Intimacy

There is a long tradition of research into intimacy and intimate relations. In sociology the more recent history of this research ranges from Anthony Giddens's (1992) understanding of the transformation of intimacy in personal relationships and Arlie Hochschild's (1983, 2012) exploration of the commodification of intimacy and the "work" of emotional labor to Carol Smart's work in her book *Personal Life* (2007). However, much of this work was developed in the context of sociological understandings of intimacy as part of the modern Western family. As Lyn Jamieson (1998) argues, multiple cultural notions of intimacy challenge Eurocentric models of intimacy whereby face-to-face (f2f) is privileged. Indeed, the question of how Asia-Pacific intimacies play out has tended to fall into the field of anthropology. For example, family and other close relationships have tended to be studied through the anthropology of kinship, which is structured through varied forms of family relationships across the Asia-Pacific and often also developed in relation to religion, as well as other cultural specificities. There is, of course, no single form or "Asia-Pacific" way of *doing intimacy*, and it is beyond the scope of this book to seek to document this diversity.

Instead, we focus on intimacy in relation to how it might be facilitated through the use of digital mobile and screen media. The same forms of digital intimacy are not being generated across the Asia-Pacific; they are always culturally and situationally specific. For instance, the work of Mirca Madianou and Daniel Miller (2013) shows how the relationships between Filipino migrants and their families left in the Philippines are mediated through what these authors call "polymedia." This concept demonstrates how intimacy is shaped through cultural, social, and technological ways of being. In contrast to the notion of the modern Western individual user of Facebook, Madianou and Miller stress how "a Filipino who uses Facebook does so not as an individual but as part of an extended family" (2013, 181). They emphasize that although local uses of digital media tend to be normative, they are normative within the contexts of local cultural and social conventions, not universal ones.

However, existing research undertaken in the Asia-Pacific suggests that digital media have certain qualities and affordances that, when activated in the context of specific sets of relationships and affects, create new possibilities for, and forms of, intimacy. In a three-year, cross-cultural case study of six locations in the Asia-Pacific, Hjorth and Arnold (2013) explored the various forms of intimate publics and mobile intimacy emerging in the region. In case studies including how social and locative media were used as a form of crisis management in post-3/11 Tokyo, generational shifts in Shanghai, political discussion and the shifting social fabric in Singapore, and the erosion of public–private and work–leisure paradigms in Melbourne, Hjorth and Arnold found that older, localized rituals around intimacy are shaping and being shaped by mobile media in the region.

Therefore in this book we do not purport to cover the question of *defining* intimacy in the context of the Asia-Pacific, and moreover we stress the need to think in terms of *intimacies* in plural. Indeed, the example of the use of mobile technologies invites us to think about intimacy differently, prompting us to ask what types of intimacy are specifically facilitated by the use of mobile technologies, as well as how these might be appropriated differently in various parts of the region by different people across a range of life stages and situations.

Digital Publics

"Publics" has been a key concept in social and humanities research for many years (for a discussion of publics and counterpublics, see, e.g., Warner 2002). We are interested in how contemporary and emergent publics form and forge themselves in relation to the presence of screen media and public art, as well as in the question of how critical public media art interventions can successfully become drivers in the constitution of critical publics.

In recent years, social media, online platforms, and mobile (as well as other digital) technologies have become part of a field of practice that might be referred to as digital activism. As John Postill points out, although activism in the region has been less attended to by digital media scholars than have protests and other forms of activism in countries such as Tunisia, Egypt, and the United States, in parts of the Asia-Pacific, digital activism is a well-established practice that was already emerging in a Web 1.0 context. Postill outlines how this included "the use of mailing lists in 1989 to protest against the Tiananmen Square massacre, the reformist movements that ushered in democratic regimes to the Philippines and Indonesia in 1998, the intensive use of mobile phones to launch mass protests against President Estrada in 2001 (Rafael 2003), and the July 2011 Bersih 2.0 rallies in Malaysia in which social media and smartphones were widely employed to campaign for democratic freedoms (Teck-Peng and Yong 2011)" (Postill 2014, n.p.). As these instances suggest, mobile media, and more recently smartphones, are part of how publics are engaged in issue-based activism and campaigning in the Asia-Pacific. They thus create certain precedents through which to think about the potential of mobile media technologies and apps to generate public engagement through arts practice. Indeed, the broader existing literature about digital activism offers us examples of the further potential of digital media to be used in critical practices that can work toward refiguring the very terms of engagement of the consumer cultures that drive their mainstream production.

Postill highlights how the digital scholar Chris Kelty uses the term "recursive public" in relation to the free software movement, itself a critical movement challenging the dominance of market-driven software production. For Kelty, Postill writes, free software constitutes a "recursive public, that is, a uniquely C21 public sphere or commons in which geeks modify and maintain the very technological conditions (or infrastructure) of their own terms of discourse and existence" (Postill 2011, 24). While we are not specifically concerned with the activities of geeks and the recursive publics that they generate through their technological practices, this example points to how critique can be mobilized to contest and refigure the tangible and invisible infrastructures and architectures of digital media.

As we discuss in the following chapters, digital and social media are already used in various political and environmental campaigns in the Asia-Pacific. When we bring these existing uses together with the more extensive ways in which digital activism has been mobilized in certain parts of the region, it becomes clear that critical mobile media arts are emerging within a space that already has a good number of existing examples of successful activist practice and online civic participation (see Postill 2011). Indeed, if we are to view ecocritical screen media and art as working toward new forms of activism in the region, then it is important to consider their potential in this domain in relation to the ways in which social, locative, and mobile media and technologies are already being used for forms of public activism across the Asia-Pacific.

In chapter 7, we discuss examples of Asia-Pacific–based participatory environmental artworks that have involved publics contributing sound recordings through digital media and thus participating in the process of making public art. The spatial and temporal elements of such projects offer us new ways to consider how publics are formed through collaboration in such digital arts events. If we think of these events of place, focused around public art, and participated in by mobile and geographically dispersed publics who are possibly living in different time zones, then the temporality and spatiality of locality become fractured into those of a place event that is not limited by linear time or geographical boundaries but still has an association with uniquely Asia-Pacific experiences, localities, and social and cultural traditions.

The Asia-Pacific already has a wealth of experience with digital activism (albeit unevenly distributed), including the examples of media arts activism and critique that we have already accounted for in this book. With this background in mind, the emergence of new publics who traverse the digital-material world as they form their critical stances on environmental issues through public art appears to be within reach. In the next section, we reflect in more depth on the implications of existing mobile media scholarship on how the engagement of such publics might be engendered and how these publics' experience might be characterized.

Intimate Publics

As we noted earlier, mobile media scholars such as Hjorth and Arnold (2013) consider one of the key characteristics of social media to be their capacity to provide various modes of visual and aural communication with greater personalization through a more intimate mode. There are potentially new possibilities for roles that mobile media can play in the modes through which publics become critically and politically engaged with environmental and climate change issues. This is because the media that make public participation possible offer, through their technological affordances and the possibilities created by apps, ways of communicating that are personalized, embodied, and emplaced and can generate a sense of intimacy and closeness. Yet they do so in a context where the same technologies and apps are engaging people in an online world that is inherently social and often public. That is, the forms of intimacy that are engendered through social media platforms often play out, and indeed are performed, in a public online domain.

The idea that forms of intimacy might be generated in contexts that are simultaneously public is not new. Writing before social media had become an integral part of everyday worlds, Lauren Berlant observed that intimacy has taken on new geographies and forms of mobility, most notably as a kind of "publicness" (1998, 281). However, in a digital-material environment, intimate relations are not simply performed in pairs or in bounded groups. Rather, they traverse the online and offline in that they are performed in physical public worlds, but also in electronic privacy (e.g., when someone privately sends a camera phone image of herself in a café to a friend), and in an electronic public that is geographically private (e.g., when we read personal messages posted to us in a publicly facing Facebook page or on Twitter while in the private space of our homes).

As Mimi Sheller puts it, "There are new modes of public-in-private and private-in-public that disrupt commonly held spatial models of these as two separate 'spheres'" (2004, 39). Advancing this idea further, Hjorth and Arnold have proposed a concept of intimate publics through which, they argue, we can understand competing histories, identities, and practices within the region. They suggest that we should understand such intimate publics as being shaped increasingly by new forms of "mobile intimacy" (Hjorth and Arnold 2013; Hjorth and Lim 2012), that is, how intimacy and our various forms of mobility (across technological, geographic, psychological, physical, and temporal differences) infuse public and private spaces through mobile media's simultaneous mediation of both intimacy and space.

These new forms of mobile intimacy therefore underpin some of the ways that public media art is able to be active in the world, that is to say, how publics become engaged with it. As we have argued, to understand how media or art can affect the world or intentionally be used to intervene in it, we need to understand its processes from the inside, and also to understand how the people who become engaged with it are engaged *within* those processes. Like the anthropologists of art Arnd Schneider and Christopher Wright, we see "creativity and meaning as something often emergent, rather than prefigured or planned" (2013, 1). Schneider and Wright call on us to attend not only to art as representation and to the intent of the artist herself or himself, but to the collaborative process through which art comes to have meaning.

By focusing on how forms of mobile intimacy can constitute the relationships between people and media art, the book's final chapters examine questions of collaboration and the generation of publics. We therefore concentrate on how forms of intimacy can be thought of as engaging publics *within* processes of critique and intervention around climate change issues. This approach offers an alternative to viewing media art as a medium for simply representing these processes to publics as if they were separate from the way they live their lives, and separate from the media technologies that they use.

Conclusion

In this chapter, we have outlined an interdisciplinary approach to understanding the relationship between art, media, climate change, and their publics. In the following chapters, we show how this is played out in the processes and experiences through which art, media, and climate change have become entangled in the region. In doing so, we ask what these can tell us about the ways in which a refigured relationship between media, art, climate change, and publics might be constituted to go beyond making a critical commentary on climate change and contribute to interventions toward new ways of living *with* and *in* the environment that are themselves ethical, responsible, and sustainable. As we propose in chapter 8, the outcome of this is to continue toward a possible or imagined world where the relationship between media, art, and climate change might be refigured.

3 Connections and Disconnections

More than half the world's ICTs are consumed and manufactured in the Asia-Pacific region, and yet there is little existing interrogation or analysis of the implications of this for either the region's natural resources or the environment. Digital technologies have increasingly become a priority globally, including for emerging economies. Postill has usefully summed up this situation, outlining how "this interest [in ICTs] has often been couched in the language of competitiveness, national security, social changing (how things are changing at present) and imminentism. A commonly held belief in economic planning circles has been that the 'Information Society' is both inevitable and imminent, and that developing nations have no choice but to embrace the new era or they will perish" (Postill 2014, n.p.).

The Asia-Pacific has been no exception. As part of twenty-first-century nation building, countries such as Japan, South Korea, China, Singapore, India, and Malaysia have all notably supported new media development through policies that are sympathetic to the building of ICT industries. A key example is how this has led to the growth of the gaming industry, particularly in Southeast Asia, and the software industry in India. In 2013, Asia accounted for 82 percent of $6 billion of global game revenue growth (Takahashi 2014). More recently, in 2013, Southeast Asia represented an estimated $661 million in revenue and was home to 130 million Internet users, among whom 85 million were gamers (*Nation* 2015). Southeast Asia is becoming recognized as "a geographic region with growing power in new media production and consumption" (Chung 2016), through which, as Hjorth's (2011) analysis implies, we might map multiple localities of divergent soft power onto national contexts. Yet while these accounts, which focus on growth, show how the mass production and consumption of ICTs have been fundamental in reformulating Asian economies and forms of national identity, the environmental consequences have been overlooked.

In this chapter, we take a cue from Maxwell and Miller's (2012) call for media studies to redress a context where the hidden wastes, inequalities, exploitation, and environmental issues associated with a growing global ICT industry have been written out of mainstream political and academic narratives of contemporary social, cultural, political, and economic processes and progress. Likewise, research into media and consumer cultures in the Asia-Pacific has burgeoned in recent decades through an increased focus on issues such as soft power (Chua 2000) and studies of lifestyle, consumption, and media (Martin, Lewis, and Sinclair 2013). However, the environmental consequences of the development of the region's economies and consumer cultures have been overlooked apart from a few texts (Lewis 2016). We respond to this agenda by taking an approach to media in the Asia-Pacific that acknowledges and examines the implicit (but not usually spoken or written about) connections between media manufacturing, production, and consumption.

First we set the scene for this discussion by accounting for how environmental issues and climate change in the region have been framed, the roles of nation-states in constituting the politics of environmental issues, and the significance of consumer culture. Then we situate screen media within this context. Finally we discuss how an alternative vision of the relationship between digital media and environment is being constituted through media and arts practice in the Asia-Pacific.

Imagining the Region through Climate Change

More than half the world's population lives in the Asia-Pacific. The region is culturally diverse, riddled with economic inequalities, and characterized by diverse but often intersecting geopolitics, economics, and cultural histories. Its nation-states are variously inflected by their encounters with colonialism, multiple and diverse indigenous cultures, and political and economic systems. Indeed, it has been suggested that the resilience of cultural diversity has made it difficult for Asian societies to respond to serious environmental challenges such as water scarcity and environmental risks to the region as a whole. For example, Brahma Chellaney has proposed:

Despite its rich history, ancient cultures and an ongoing economic renaissance, Asia is the only continent other than Africa where regional integration has yet to take hold. In fact Asia's political and cultural diversity has acted as a barrier to collaboration and integration. Consequently, Asia lacks institutions to avert or manage conflict, even as greater prosperity and rising nationalism are stoking territorial and resource disputes. (2012, 154–155)

Like Chellaney (2012, 2013a, 2013b), many see water as an increasingly scarce and contested resource that will inevitably generate widespread geopolitical conflict (Hoekstra and Chapagain 2008; Shiva 2002; Wirsing, Stoll, and Jasparro 2013). Whether or not such conflict will actually be a widespread outcome is unknown, but existing examples demonstrate how local lives and places can violently be disrupted by state policies designed to benefit a nation as a whole. For instance, in 2011 the *BBC News* reported that since 1997 more than 1.4 million people had been involuntarily displaced by the Three Gorges Dam hydroelectricity project on the Yangtze River in western China (Dawson and Farber 2012, 40). Completed in 2006, this massive dam also created unforeseen environmental and ecological problems that, with the benefit of hindsight, the Chinese government has since publicly acknowledged (Bo-Jie Fu et al. 2010; Bristow 2011; Hvistendahl 2008; Maczulak 2010).

Although the significant environmental costs of dams are now better understood than when many were constructed in the twentieth century, hydropower is nonetheless a relatively clean and cheap form of energy, and there are plans for many more dams in China and the Chinese territory of Tibet. Meanwhile India, Nepal, Bhutan, and Pakistan also seek to harness the river system of the Himalayas. Other countries in the region, such as Laos, Cambodia, Thailand, and Korea, are planning new dams. Even in Australia, the government was considering a new dam-building scheme despite recurring problems with drought (Cullen 2013). Hence across the region the aim to secure reliable sources of energy is a significant factor for (albeit uneven processes of) national development. Moreover, such processes are becoming increasingly competitive as each nation-state depletes more natural resources with little consideration for the region as a whole.

Technological changes to the life of rivers, along with the human and nonhuman ecologies that depend on them, are symptomatic of deeper currents of change in the Asia-Pacific. Of these, three broad and interwoven processes stand out. The first is clearly the uneven processes of modernization, manifested especially in the extraordinary speed of urbanization across the region, combined with the rapid technological change that has been so prolific in the megacities of East Asia. The second is an inexorable increase in regional urban populations, known as "megacities," and the intensifying pressure they place on the supply of energy, water, and rural resources. This phenomenon is making resource scarcity a part of the everyday lives of ordinary people and thus generating experiential forms of awareness relating to climate change. Third, processes of change are unfolding in response to the gradual emergence of environmental and ecological deterioration of the region exacerbated by global climate change.

In chapter 1, we pointed to the scientific consensus that the warming of the global climate system is "unequivocal" and that anthropogenically caused rising global temperatures will "very likely" result in rises in sea level and generally hotter weather (IPCC 2013). The report also suggests that, by the late twenty-first century, an increase in tropical cyclone activity is "more likely than not," particularly in the Western North Pacific (IPCC 2013, 5). In this context, we can therefore gain a sense of how climate change is emerging as experienced through extreme weather conditions. Moreover, because it is coupled with urban population growth that in some cases exceeds infrastructural capacities, climate change is also experienced as resource scarcity in everyday life.

The Asia-Pacific has recently been affected by a switch in the natural Southern Oscillation Index from dry El Niño conditions to the cooler and wetter conditions of La Niña, which in turn are more conducive to heavy rains, floods, and cyclones. Previously, official media reports rarely speculated on how an intensification of these natural conditions might be connected to global climate change. More recently, however, there has been a shift toward a greater public acknowledgment that human activity might be contributing to more extreme weather events. In 2011, for example, after severe flooding and the category 5 Cyclone Yasi had crossed the Queensland coast, the Australian scientist David Karoly was widely reported in the national press for his commentary connecting the extremity of events with global climate change. As he remarked, quoted by Bridie Smith (2011) in the *Age* newspaper: "Australia has been known for more than 100 years as a land of droughts and flooding rains, but what climate change means is Australia becomes a land of more droughts and worse flooding rains."

With the rise in global ocean temperatures, the ocean has more potential to generate powerful tropical winds and cyclones (Elsner, Kossin, and Jagger 2008), and although the direct outcomes of such conversions cannot be proved conclusively, people are nonetheless beginning to infer a causal link between extreme storms and their anthropogenic causes. This was evident in the example of the category 5 super-typhoon Haiyan in November 2013. Known as Typhoon Yolanda in the Philippines, where it killed more than six thousand people, the typhoon was one of the strongest storms ever recorded. Although the Philippines was the worst affected area, the typhoon also caused significant damage in China, Vietnam, and Taiwan. Indeed, for the sociologist Tarique Niazi, the relationship between climate change and extreme weather is clear:

The Philippines is among the Asian nations that seem to have become ground zero for climate change. Many coastal and island nations in Asia are already among its fellow sufferers. In the Bay of Bengal, Bangladesh has become the most exposed country to worsening climatic events. Year after year, it is battered by cyclones of ever higher intensity and ever greater frequency. In a single event of extreme weather, hundreds, and sometimes thousands, lose their lives. Besides, economic and social dislocation visits upon the millions, leaving them stranded for months, and even years. If global mean warming exceeds 1.5 degree Celsius, the largest chunk of coastal Bangladesh will begin to teem with "climate refugees." (2013, n.p.)

This narrative also has a clear culprit. For instance, Kerry Emanuel, an IPCC member and atmospheric scientist at MIT, argues that developing nations like the Philippines "are suffering for the sins of developed countries that followed the path of carbon-heavy development" (Niazi 2013, n.p.).

Indeed, this discourse prevails across some Philippine interpretations. In the wake of Typhoon Yolanda, the Filipino diplomat Yeb Sano gained world media attention at the UN Climate Change Conference in Warsaw with his emotional televised appeal to the world to reduce global carbon emissions. After going on a hunger strike, Sano developed an online petition that has subsequently received more than three-quarters of a million signatures. Even if the cautious findings of the latest IPCC report suggest Sano's appeal is premature in the way it blames anthropogenic carbon emissions as the hidden human hand behind natural events, his appeal is nevertheless significant as a

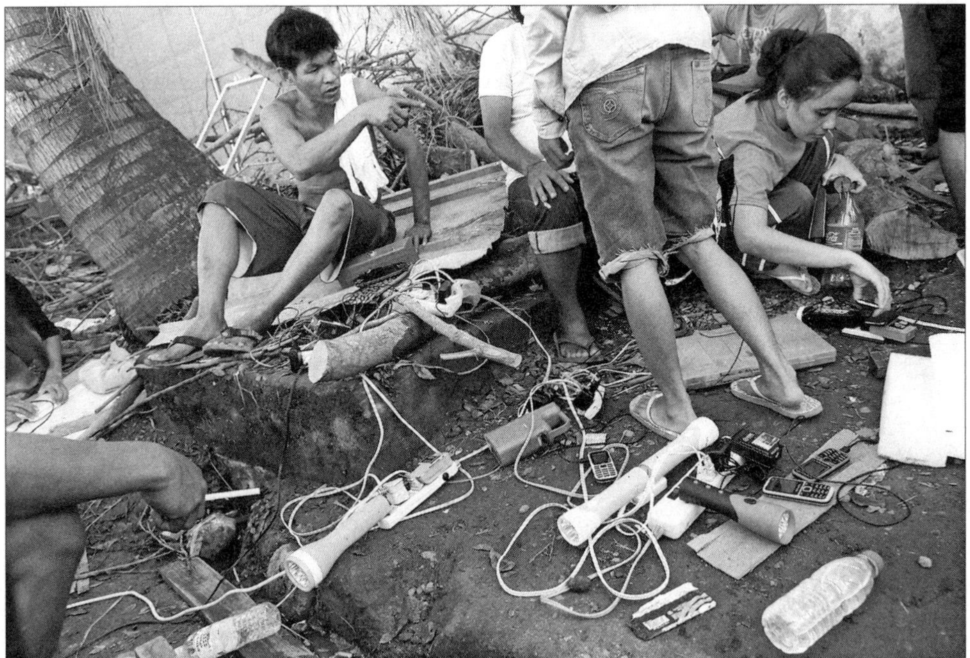

Figure 3.1
The material realities of mobile phone use in the Philippines after Typhoon Haiyan, 2013. Photo: AP via AAP/Vincent Yu.

global media event connecting local experiences and regional conditions to the more general shift in the global climate. Sano's website proclaims that "new realities require new politics" (AVAAZ.org), and although the irony that media technologies were so central to his campaign should not be missed here, the call for a new approach to the problem is endorsed by the sociological and scientific analyses noted earlier.

If we look at Australia, where the relationship between national infrastructures, weather, and human experience is differently constituted, the example of Cyclone Yasi in 2011 shows us how thousands of Australians sought refuge in urban shopping malls and could rely on well-equipped state emergency services. By contrast, Sano's appeal to the world from the position of a developing country such as the Philippines was a salient reminder of the millions of people in the region who were most vulnerable, as those who "contribute least to global warming are both facing the most severe consequences and have the least capacity to cope" (Burgmann and Baer 2012, 3).

However, while recognition and awareness of these environmental issues across the region and beyond are increasing, responses to them are often shaped by political and economic fields of activity that are influenced by competing agendas. In the next sections, we consider two influential strands of this configuration: the question of cultural nationalism, and the deeper problem of a capitalist growth economy. These two broad processes are not mutually exclusive, as is clear in the ways that official appeals for national security are so often measured in response to how multinational interests continue to dominate fluctuating global markets. Hence, for example, in a relatively buoyant economy such as Australia's, the government has cited the global financial crisis of 2008 as a reason why the nation-state must continue to secure its own financial stability ahead of its commitments to foreign aid and must prioritize foreign investments in Australian extractive mining before contributing to international efforts to mitigate climate change.

Contesting the Nature of Localities

From one perspective, the Asia-Pacific has been conceptualized as a series of contested satellite localities with different experiences of colonialisms, imperialisms, and soft power flows. In contrast, notions like "Asia as a method," put forward by Kuan-Hsing Chen (2010), have suggested new ways in which the region's contested localities can be framed through a process of de-imperialization and a reading from within (as opposed to through imposed Western constructs). Within these emergent narratives, ideas about nation-state boundaries and flows have been revised. This approach generates insights into how climate change issues such as those outlined at international levels by the IPCC are (or are not) implemented at the level of local and national policies.

Nationalist values are manifested in a range of ways across the Asia-Pacific, particularly in economically developed nation-states such as Korea, Australia, China, and Japan. Postwar Japan, for instance, was characterized by a general cautiousness toward revivalist cultural nationalism that was associated with historical cultural formations and conservative politics (Yoshinko 1992), yet certain cultural traditions are still maintained, presenting a sharp contrast to Japanese popular culture. Transnational versions of popular culture and modes of consumption have increasingly become sites where national identities are contested (Iwabuchi and Chua 2009). However, there are simultaneously tensions between nation-states in the region. Along with long-held political tensions, such as those between China and Japan or North Korea and South Korea, disputes have emerged over ecological questions that recur in both official regional media reports and social media. These examples show state-led nationalism being contested by environmental and animal rights activists.

For example, Sea Shepherd, an Australian-based environmental group, was formed in 1977. Its mission is "to end the destruction of habitat and slaughter of wildlife in the world's oceans in order to conserve and protect ecosystems and species" (Sea Shepherd 2014). The group uses direct-action methods to achieve these ends. When the group filmed its dangerous encounters with Japanese whaling ships in the Southern Ocean, the footage was reinterpreted by news media as an example of national conflict between Japan and Australia (Merlen 2014). Undercover filming by the animal rights activists Animals Australia has also received significant mainstream media exposure in Australia and recently led to a change in government policies over the conditions of live cattle exports to Indonesia (Animals Australia 2014).

In contrast with the call for action on regional and global environmental issues via social media campaigns by groups such as Get Up, Animals Australia, and Sea Shepherd, the last two decades have also seen a resurgence of cultural nationalism. This has occurred in Australian policy and in some domains of everyday culture, such as popular celebrations of Australia Day and a renewed focus on the Anzac legends memorializing Australian engagement in World War I, which served to define Australia as a modern nation-state. This cultural nationalism is also developing in a context where there is a related new sensitivity about national borders that is leading to the increasing exclusion of refugees arriving by boat. Although this notion is contested by activists, it also has populist support. This revival of cultural nationalism in Australia therefore has both human and wider environmental implications, since successive governments have prioritized national vested interests in business and mining over effective policies on carbon emissions (Burgmann and Baer 2012). There is, however, a growing public awareness in Australia of issues relating to both refugees and environment. This greater awareness is contributing to the production of a series of screen media and art interventions that are continuing to emerge at the time of our writing this book.

In China, with the erosion of traditional cultural values after the Cultural Revolution of the sixties and seventies, the "soft power" of cultural nationalism has been encouraged by China's political elite—especially since the protests in Tiananmen Square in 1989 (Associated Press 2014). In contrast to the iconic image of "Tank Man" of June 4, 1989, that appeared on many screens globally (which we discuss later), the dazzling screen spectacle of the opening of the 2008 Summer Olympic Games in Beijing represented a shift in the global currency of images of Chinese cultural nationalism. The Beijing Games, along with their associated urban development policies and massive new building programs, were also well aligned with the aims of China's political elite (the Communist Party of China, or CPC), who aspire to an extensive national agenda of heroic modernization. In 2013, however, on the twenty-fourth anniversary

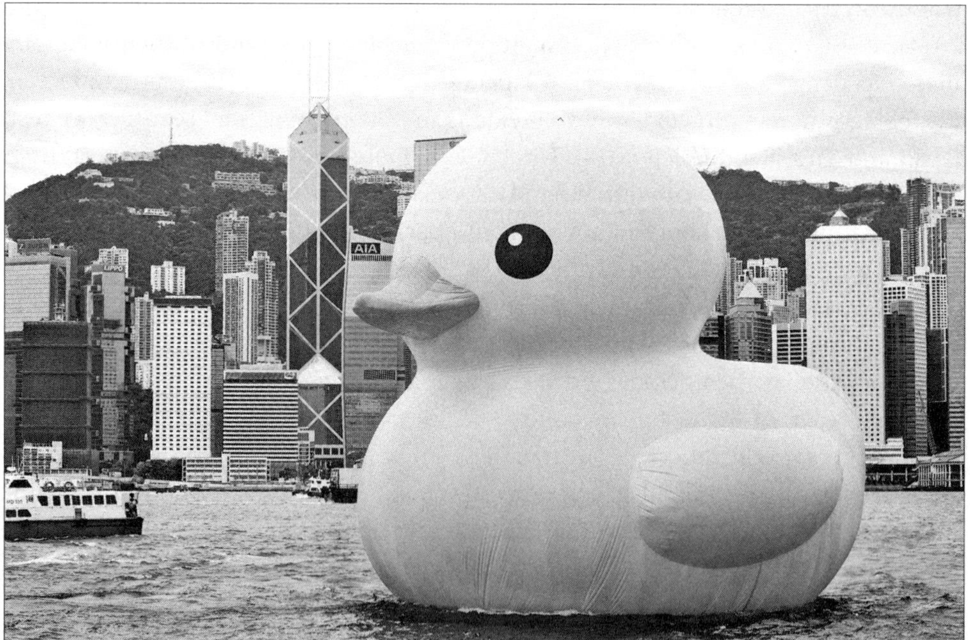

Figure 3.2
Florentijn Hofman, *Yellow Duck*, 2013. Courtesy of the artist.

of the events in Tiananmen Square, tens of thousands of people gathered in a park in central Hong Kong to protest against the hegemony of the CPC (Mullany and Buckley 2013). A month earlier, an unexpected visitor—a giant yellow duck—had floated into Hong Kong harbor.

An international artwork by the Dutch artist Florentijn Hofman, *Yellow Duck* has traveled around the world, appearing in many major ports. According to Hofman, the duck "knows no frontiers, it doesn't discriminate people and doesn't have a political connotation. The friendly, floating Rubber Duck has healing properties: it can relieve mondial tensions as well as define them. The rubber duck is soft, friendly and suitable for all ages" (Hofman 2013, n.p.). In June 2013, as crowds gathered in the Hong Kong protests, Sina Weibo (China's most popular social media site) blocked access to information about the anniversary of the event, including censoring the term "big yellow duck." The censorship was designed to prevent the circulation of a Photoshopped version of the iconic "Tank Man" image with the military tanks replaced by a row of smiling rubber ducks (Kaiman 2013).

A Hong Kong designer, Michael Miller Yu, came up with a montage that, according to Sina Weibo, quickly went viral across China before being banned (RNW 2013). While the media adaptation of *Yellow Duck* does not focus on environmental issues as such, it makes a nonconfrontational point of dispelling the imagery of aggressive nationalism, which in the longer term offers an alternative path to approaching regional environmental issues. Yet this context is more complex because the environment, nature, and risks relating to these protests are not only imagined and contested by government and artists; they are also imagined through corporate interests and manifested in the emergence of specific consumer cultures.

We continue this discussion by reflecting on the example of China, which we have selected specifically because it demonstrates how the ways in which the environment is imagined can become bound up with emergent middle-class consumer cultures. ICTs have become an essential element of these cultures, given that ICTs are elements through which relationships between consumption, production, power, and labor are entangled. China is no exception. In China, various scandals related to contaminated natural products periodically cover the front pages of newspapers. Simultaneously, populist films (such as those by Feng Xiaogang) present nature as "an image to be appreciated or as a symbol of national grandeur rather than as a treasure that is to be protected" (Neri 2013, 122). As Corrado Neri points out, "Nature is a selling tool for all kinds of products, designed ultimately for upper class leisure" (122). Here nature is being mediated as a sign of luxury and, significantly, as a desirable alternative to urban density. This appropriation of the notion of nature is not limited to China; it is common to Asian megacities, where only the extremely wealthy can afford a garden or easy access to rural space as a site of leisure.

Indeed, in this context, the mediation of nature takes on representational forms that could be interpreted as calling on intimate human desires and longings and are designed to appeal to individual aspirations. Such mediations may also be culturally engaged with nationalism to the extent that people share a sense of pride in the beauty of what Neri describes as "national grandeur" (2013, 122). They are part of a growing middle-class imaginary in which culturally specific, shared aspirations bring together everyday media contexts with experiential domains where the impact of climate change is "felt" or evaded as people navigate differential access to things like water, energy infrastructures, and air-conditioning.

Such forms of cultural nationalism and consumer culture appear to go hand in hand. As Robyn Eckersley has proposed, "only the state can tackle the ecological myopia of the market" (2004, 82), but since nation-states are committed to growth economies in ways that are bound to international capital, these are the crucial "contradictory imperatives" that prevent decisive action on the mitigation of climate change (102). However, as we demonstrate later with examples of engagement through screen media and public art, when environmental and climate change issues are repositioned by artists or activists, they make visible forms of resistance that contest aspects of consumer culture across national and transnational realms. This reconfiguration of the role of climate change is deeply embedded within screen media and art practice in the Asia-Pacific. The potential of these media to be used in ways that transgress the national and the interests of cultural nationalism correlates with what Michael Curtin calls "media capital" in the region whereby, Curtin argues, contemporary media flows "do not necessarily correspond to the geography, interests or policies of particular nation-states" (2003, 204).

Reimagining the Environment in an Age of Big Data

In the first part of the chapter, we outlined how the Asia-Pacific has already been imagined in relation to environment and climate change. Yet these accounts do not attend analytically to the ways in which media are implicated. In this section, we first discuss the connections between media and climate change that have been made in the existing literature before suggesting how they enable us to reenvision the question of environmental issues in the Asia-Pacific.

As Maxwell and Miller observe, the "invidious growth" of ICTs and consumer electronics in the global North (Australasia, western Europe, Japan, and the United States) sees millions of tons of electronic and electric waste (e-waste) being dumped annually, predominantly in the global South (Latin America, Africa, eastern Europe, South and Southeast Asia, and China) (2012, 3). This situation is changing, however, with China and India increasingly "generating their own deadly media detritus" (3) as growing centers for ICT manufacturing and consumption. Maxwell and Miller suggest that the pervasiveness of the technological sublime—"a totemic, quasi-sacred power that industrial societies have ascribed to modern machinery and engineering" (4) after World War II—has hidden much of the connection between media technologies and ecological decline. This situation is no less prevalent in the Asia-Pacific, with its unprecedented rise in technologically savvy middle-class consumer cultures (Chua 2000; Robison and Goodman 1996), as well as the increasing use of mobile technologies in small-scale businesses and other enterprises.

To some extent, the relationship between global issues and media in Asia-Pacific countries is made explicit through mainstream news and campaigns redeploying media content. For example, a consideration of iPhone manufacture easily invokes images of exploited Chinese workers in Foxconn factories (Qiu 2012), as well as similar incidences in other localities in the region. Less frequently discussed is how, as the middle classes grow, Asia-Pacific consumers are also participating in creating a demand for the continuous production of new but soon-to-be-obsolete mobile media technologies. Moreover, it is not only the impact on the environment of the hardware of mobile phones and other technologies—both through their manufacture and as waste or pollution—that is of concern. In addition, as Big Data becomes an increasingly central element of everyday personal and corporate lives and activity, questions about data storage and e-waste are also magnified. It is easy for people to forget that the storage of immaterial or "digital" data has material impact. The digital needs physical infrastructure such as cables and physical buildings to store it. For instance, the ironies of the relationship between data, e-waste, and the environment are demonstrated in discussions of the digital visualization of environmental crisis. Allison Carruth and Robert P. Marzec point out how, with "the swift dissemination of contemporary [digital] visual culture, it is easy to take images of environmental crises and environmental movements for granted" (2014, 206). Their argument suggests that processes of image making and visualization have become naturalized in "twenty-first-century ecological realities" (Carruth and Marzec 2014, 207). They describe a global context increasingly defined by digital imagery:

From climate change models to [the] World Wildlife Fund (WWF) "Species Tracker," environmental visualizations are political and politicized as much as aesthetic and aestheticized. Digital infrastructures (like the data centers that run the cloud and the global positioning systems [GPS] that generate high-resolution images of the earth's surfaces) and digital media (including computer-generated imagery in filmmaking and viral online videos) have come to color our sight lines perhaps even more extensively than industrial technologies defined the landscapes of the nineteenth and twentieth centuries. (Carruth and Marzec 2014, 207)

Once explained, the ironies seem obvious. Carruth and Marzec claim that these "acts of visualizing environmental crises and environmental solutions often hide, moreover, the ecological footprint of image-making technologies themselves, which tend to promise the transparent representation of empirical facts or shared experiences" (2014, 207). Unraveled as such, the relationship between screen media and environmental issues is part of the same problem that Maxwell and Miller (2012) identify in showing how ICTs more widely have been shrouded in a technological sublime or "enchantment" that has clouded our ability to recognize the impact of e-waste.

In the case of visualizations of the environment, Carruth and Marzec connect these ways of depicting nature in climate change images with the historical trajectory that Mirzoeff has referred to as part of "anthropocene visuality" that "obscures rather than reveals both environmental and social injustices" (Carruth and Marzec 2014, 209). According to Mirzoeff, anthropocene visual culture can be linked to early forms of visualization rooted in the perspectives of eighteenth-century military theorists, suggesting that within the West, the "conquest of nature has become integrated into Western aesthetics throughout the Anthropocene" (2014, 219).

We are wary of whom the "we" of the "our" used by Carruth and Marzec might refer to, precisely because such accounts of what images "tell" us do not account for what people do or imagine with images, or for the different localities in which they are engaged (with). The matter of how, in a world of representations, "the environment" is depicted, defined, objectified, and debated is important for our analysis. However, rather than seeking to posit a generalized analysis of how such images can be read by an elusive "we," we focus instead on how those who think they can read such images might comment on them. In particular, we focus on how alternative forms of visualization have been developed as a critical response or as part of a different trajectory to the work with which scholars of visual culture normally identify. Indeed, Mirzoeff's genealogy of visual culture is overtly Western, neglecting the culturally specific histories of visuality in the Asia-Pacific and how these have become interwoven with the ways in which contemporary Asia-Pacific digital visual cultures of climate change are composed and interpreted.

Analyses such Mirzoeff's need to be understood precisely as the work of modern Western cultural theory, and as such they do useful work in revealing how certain dominant ideologies about the world are visualized. However, there is another way to think about how the relationship between visual cultures and screen media is constituted. This, as we outlined in chapter 2, drawing on the work of Ingold, involves understanding images not as being *of* the world but as being *in the world* with us and a part of the environments that we inhabit. Following Ingold, this means that we might understand images in terms of what people do and imagine *with* them (Ingold 2010), and not simply as imposing on us powerful, a priori discourses that expound the values of institutions and global capital.

This situation is particularly applicable for understanding how UCC and UGC (user-generated content) are contributing to how climate change is visualized and how it is commented on (through words and images) through collective projects. As we discuss later, these visualizations and commentaries take place via social media platforms and in the context of participatory public art platforms and events developed in the Asia-Pacific. The Asia-Pacific example is one where art, media, and environment become entangled with respect to the trajectories of local, regional, and global visual cultures and the ideologies that inform them. However, this complex relationship cannot be understood without also examining the specific activities that artists and publics are involved in.

Reconnecting Art, Media, and the Environment

In recent years, the relationship between art and environmental issues has become increasingly explicit, both in a general sense and in its specific development in the Asia-Pacific. For instance, the writer Andrew Brown, in his 2014 review of the relationship between art and ecology, suggests that environmental issues have "become part of the artistic mainstream." This means that art and ecology have become part of a world of frequent "international exhibitions, conferences and festivals on ecological themes," and "artists of all kinds are being commissioned in ever greater numbers to explore humankind's impact on the planet" (Brown 2014, 5). Brown argues, "From being a peripheral activity, art that seeks to ask searching questions about the environment is now firmly center stage, at once responding to and shaping debates in a broader society" (6). This is no less so in the Asia-Pacific, where over the past decade, artists, activists, and academics have begun to question the relationship between art, media, and the environment.

A brief review of some recent Asia-Pacific projects shows how an increasing number of artists are critically exploring the development of a visual culture that goes beyond objectifying the environment. Artists are playfully providing alternative pathways for imagining and engaging with the environment, and media are playing a crucial role in this. In Australia, for instance, forms of creative practice that robustly engage with the processes and materials that stand for dominant ideologies and flows of capital and their environmental implications are cohering into a vibrant community of critical practice. For example, the Australian project *Spatial Dialogues* sought to collaborate with partners in Shanghai and Tokyo to make artwork interventions into the urban and media everyday. This project is discussed in detail in chapter 7. Another example is the Environmental Research Initiative for Art (ERIA), which has also sought to use art and new media to intervene in public debates about climate change. ERIA, for example, aims to "trial renewable energy with analogue and digital media technologies to create innovative models of public art that are environmentally sensitive and self-sustaining. ERIA aims to conceive, construct and install site-specific public art that accommodates 'eco-logical' practices in ways that regenerate physical environments, while also producing new and meaningful exchanges with a community of users" (2014, n.p.).

The Australian academic Jill Bennett has linked debates about sustainability and media through art interventions. In her *Curating Cities* project (including a database launched in 2013), artworks explored themes through works with titles such as *Atmosphere, Energy, Renewal & Regeneration*, and *Waste, Recycling, Consumption*. The Australian artist Lauren Berkowitz transformed a gallery space into a "temporary greenhouse" (Brown 2014, 239) to reflect on sustainability in relation to nonnative species that are "destroying the fragile Australian continent since colonization. At the same time, she proposes indigenous and succulent plants as a way of broadening our choices to include more sustainable food sources" (O'Brien in Brown 2014, 239). As Brown observes about Berkowitz's work:

Lauren Berkowitz works with living and ephemeral materials to evoke the passage of time and the cycles of life and death. She also often makes use of found and recycled objects to address the degradation of the natural environment caused by human activity, frequently utilizing some of the vast quantity of plastic bottles, pots and packaging that we generate and dispose of every day. In her view, this huge excess of human-made waste is just one of the ways we have lost touch with nature and any sense of where and how our food is produced. (Brown 2014, 239)

In the Philippines, the photographer MM Yu appropriates the material culture of corporate capitalism, in this case waste. She uses it critically, taking pictures of orphaned objects found in everyday Manila life. Her pictures are poetic gestures toward the rapid consumption and poverty that one encounters in the streets of Manila. A pile of mass rubbish is rendered as a colorful media landscape. Its familiarity is transformed into a terrible sublime. Images of the Philippines, a developing country that regularly suffers environmental disasters such as typhoons, are often held up in global media as a repository of climate change. In contrast, developed countries in the region, such as

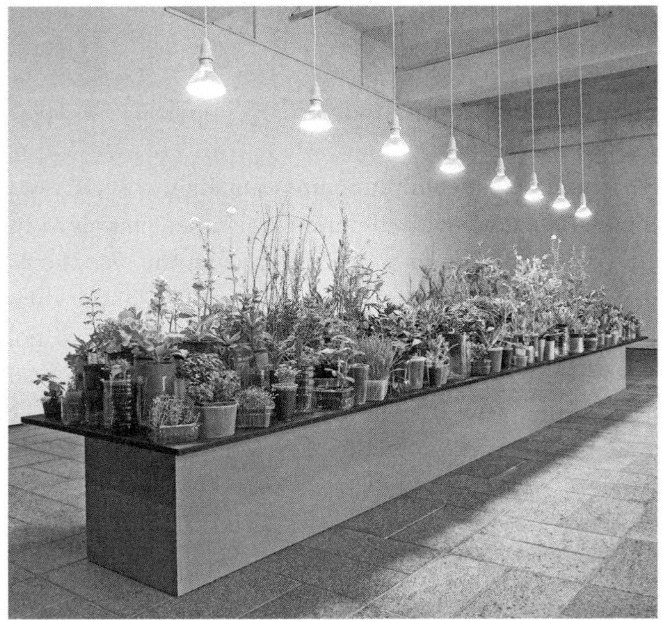

Figure 3.3
Lauren Berkowitz, *Mana*, 2013. Photo courtesy of the artist and La Trobe University Museum of Art, Melbourne.

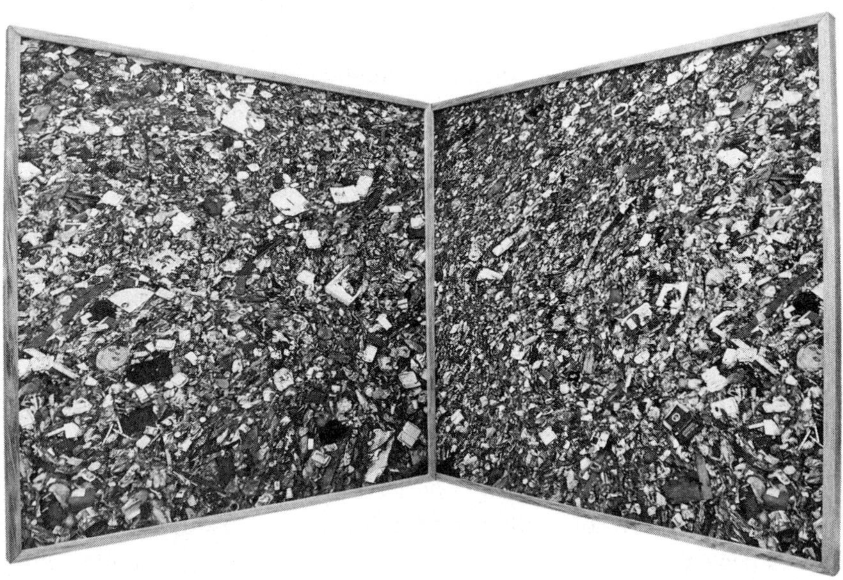

Figure 3.4
MM Yu, *A Few of My Favorite Things*, 2009. Mixed media. Courtesy of the artist.

Singapore, often present images of themselves as environmentally responsible and possessing a sustainable-looking urbanism.

Whereas the critical perspectives on visual cultures of environmental crisis that we discussed earlier suggest that such visualization "obscures rather than reveals both environmental and social injustices" (Carruth and Marzec 2014, 209), the Asia-Pacific art projects we have discussed are positioned rather differently in relation to the environment. They place the issue of environment at the center of the artwork by incorporating the very artifacts that stand for the processes and outcomes of environmental degradation. Lauren Berkowitz (fig. 3.3) brings the plastic bottles that are part of the "problem" into her work, but repositions them for an afterlife whereby they can help plants to grow rather than destroying nature.

MM Yu (fig. 3.4) likewise creates her art from and through waste, using the artifacts of climate change to make explicit her critique. These examples thus begin to suggest alternative ways not only of accounting for climate change but also of beginning to engage with it. They approach climate change not as something that is external to the world that we live in and are part of but as something that is part of the environment we inhabit. In these Asia-Pacific–based works, therefore, we do not see images that seek to "capture" climate change or represent an environment in crisis. Rather, we see a deeper and more critical interrogation of the processes through which climate change is happening.

Such approaches are particularly well demonstrated by looking at how art has been engaged as a critical response to the place of bottled water in the Asia-Pacific. Bottled water has to some extent become defined in the region as a luxury item that connects consumers with "nature" (Hawkins 2005). Yet for many middle-class urban dwellers, it has also come to be perceived as a necessity in the face of a lack of readily available drinking water from existing water supplies and infrastructures. However, the production of plastic bottles creates pollution at every stage of manufacture, from the oil used in producing plastic to its lack of biodegradability when discarded.

As the Australian cultural studies scholar Gay Hawkins has observed, "Far beyond terms like rubbish, trash, or litter, the idea of waste can provoke a minefield of emotions and moral anxieties" (2005, 1). In her recent collaborative work, Hawkins explores the politics of bottled water beyond the usual political or environmental tropes to consider the ontological dimensions of drinking bottled water as embedded within many everyday practices (Hawkins, Potter, and Race 2015). Though recyclable, most bottles end up in landfills or, more notoriously, in one of the five great ocean garbage patches. In his *One Second (Plastic Water Bottles 5982)* (fig. 3.5), the Australian artist Stephen

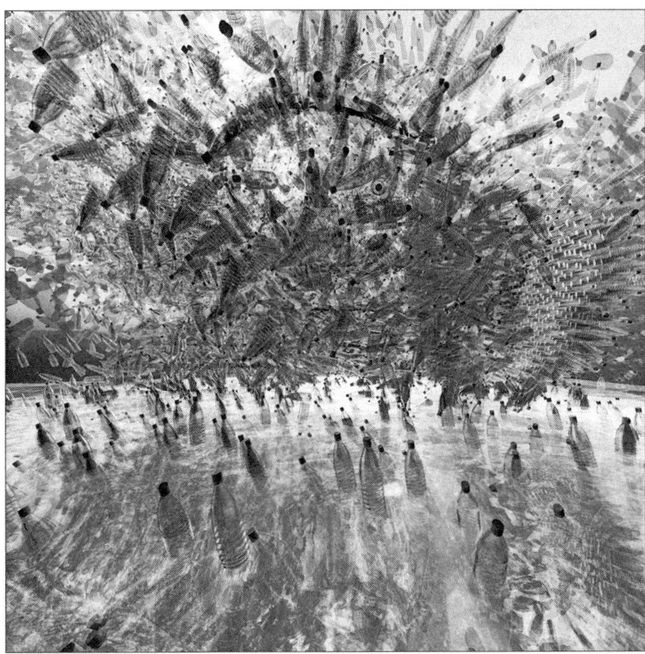

Figure 3.5
Stephen Haley, *One Second (Plastic Water Bottles 5982)*, 2010. Lightjet photograph. Courtesy of the artist.

Haley represents the pollution of the ocean in a digital print that illustrates the global scale of commodities at the level of human production and consumption.

In his series *One Second More* (see fig. 3.5), Haley amassed a range of commodities as they were produced globally in one second in 2010. The global production of plastic water bottles has increased since 2010, when, as Haley shows, 5,982 plastic water bottles were produced in one second. And despite the seductive submarine colors of his work, at this rate of production and consumption, the density of the amassed bottles suggests an expansive spatial field that will eventually become inundated with plastic. These projects, particularly those focusing on plastic and waste, demonstrate how public art, as it critically approaches environmental issues, has a powerful potential to address the "hidden" elements of otherwise often conspicuously luxurious and desired consumer objects. These artworks simultaneously imply an excess of consumer culture and invoke the disastrous consequences of the damage that the manufacturing and disposal of these products generate.

The projects discussed in this section were produced across different countries in the region and do not necessarily have direct relationships with one another. Yet collectively they begin to offer us a rather different image of the relationship between the Asia-Pacific and climate change from those we have discussed earlier. This different image emerges from political and institutional discourses, from critical analysis of visual cultures of environmental crisis, and from critical accounts of consumerism and their role in reimaging national and transnational practices.

Conclusion

In this chapter, we have outlined how the Asia-Pacific might be reimagined through the prism of climate change, and highlighted how media and arts practice are becoming implicated in how climate and nature are defined, contested, and debated. We have explored some of the environmental issues confronted in different parts of the region, as well as common themes in environmental questions. We have also shown how cultural nationalism and the emergence of new consumer cultures are part of the ways that climate change realities, policies, and notions of the environment are shaped, contested, and experienced.

Screen media and arts practices are playing an important role in the constitution of consumer cultures. They contribute to forms of contestation and activism that bring powerful critiques of both cultural nationalist policies and certain forms of commodification into the public domain. In the next two chapters, we advance this exploration further by focusing on how mobile media and art platforms are implicated in critical and advocacy roles in relation to environmental issues, and how these are experienced in everyday life.

In this chapter, we explore how mobile media are being engaged to visually represent the environment in the Asia-Pacific. We do this through a particular focus on the visualization of disasters. Environmental disasters, which have been connected to climate change in the region, have been well documented, although existing accounts of them attend little to the ways in which mobile media and arts practice have become connected to them. We examine five aspects of the phenomenon of the environmental disaster: remediation practices; the unevenness of relationships between mobile media and environmental disasters in the region; how mobile media are being used in relation to climate change and environmental crisis through the work of selected artists; the tension between amateur and professional media art; and how camera phones as embedded within emergent visualities affect how we understand the environment from within as we move around in it.

Our discussion is informed by three key themes: environmental costs, user-created content (UCC), and consumer cultures. First, mobile screen media are implicated in the environmental costs associated with their manufacture and the production of e-waste, as are digital media technologies more broadly. This background informs our discussion of the ways that they are used in relation to issues or communications around climate change and notions of environmental crisis. Second, much research into mobile media use has focused on the notion of UCC, tending to characterize the "amateur" everyday user as a hybrid producer and user, such as through the concept of the "produser" (Bruns 2005). This field of research has also developed a strong focus on camera phone photography (e.g., Hjorth and Pink 2014), which has included attention to Big Data visualizations that often appropriate mobile media information, including GPS data, to make world maps and camera phone apps such as Instagram with its retro filters fulfilling a "nostalgia for the present" (Jurgenson 2011). This body of literature offers an important basis for understanding how everyday users engage with mobile media, as well as outlining the user activities and types of software and apps that contribute to the production of e-waste.

Third, existing research into the uses of mobile media in the Asia-Pacific shows how they are part of the rise of consumer cultures that are playing a detrimental role in processes associated with climate change. We account for this while at the same time acknowledging that they are accompanied by uses that contest these consumption practices. These uses include everyday consumers rejecting the impulse to always have the latest model, the use of mobile media among political activists and campaigners, and uses of mobile media in other forms of democratic civic participation, as well as in development and aid programs.

Remediation Practices in Asia-Pacific Art

Remediation practices form part of the way that digital media are engaged in contemporary art in the Asia-Pacific and beyond. They involve artists bringing together, combining, and remixing old and new media into new configurations. In the context of using remediation in art, this can also mean appropriating media into art forms. Indonesia offers a good number of examples of artists using new and remediated media to rethink the relationship between community, sustainability, and the urban. Established in the 1990s with President Suharto's totalitarian New Order regime (1967–1998) as the backdrop, the artist collective Apotik Komik (the Comics Chemists) in Yogyakarta deploys popular media such as comics to provide political and social commentary on everyday life (Jurriens 2014). The Australian conceptual artist and musician Danius Kesminas has a long-standing collaboration with the punk DIY (do-it-yourself) band Punkasila. Their songs and costumes toy with Indonesian history and popular culture to humorously comment on politics.

Mobile media bring our attention not only to debates around mobility but also to the role the social plays in art. Kester's notion of "dialogical aesthetics," as discussed in chapter 1, offers a way to understand how relationships between art, media, and the environment are being shaped through and generate social relationships. As Kester suggests, through the entanglement of dialogue and interaction, the politics of aesthetics help us to locate the various tactics and methods being used by artists in the region.

A key example of how social media are becoming engaged in this relationship is the work of the Saigon-based Propeller Group, comprising Tuan Andrew Nguyen,

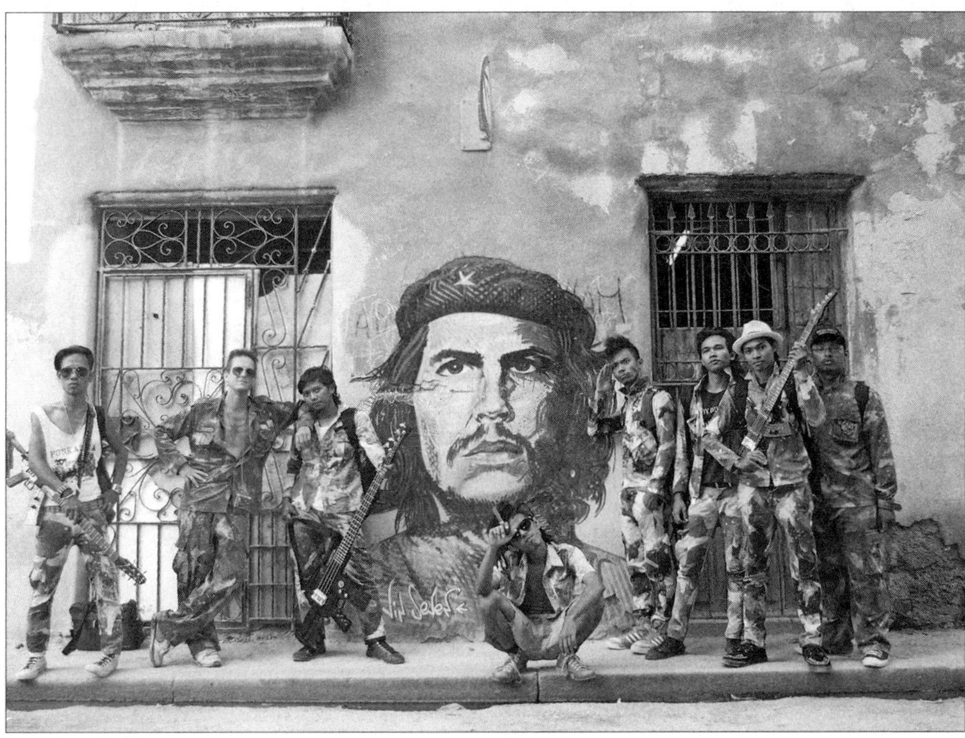

Figure 4.1
Punkasila, *La Misión a Cuba del Rock Combativo*, Tenth Havana Biennial, 2009. Photo: Reynier Rodriguez Vazquez.

Phunam, and Matt Lucero. The Propeller Group has an ongoing project called *Viet Nam: The World Tour* that "uniquely challenges the content of social media as not only a high-art form but also an archival database that can refresh the online searchable terms prescribed to theories of cultural identity and nationhood" (Butt 2014, 100). In locations like Vietnam, where online content is highly censored by the government, art practices that intervene online are as social as they are political. Increasingly, screen culture has become synonymous with the use of social, mobile media as a vehicle for political agency. So, too, artists and the general community deploy screen media as a way to speak about the environment and emergent agencies and politics.

Mobile media and mobility are key features of the work of the Indian artist Gigi Scaria. Scaria's work views contemporary urban scapes in Asia as creating their own hybrid landscapes that are at the same time local and global, familiar and strange.

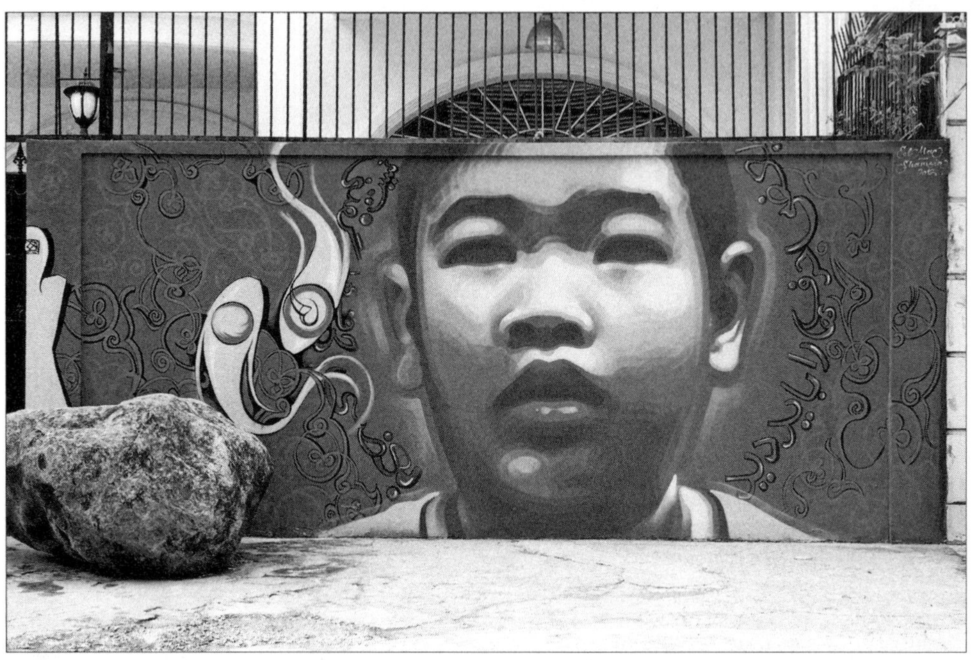

Figure 4.2
The Saigon-based Propeller Group, *Viet Nam: The World Tour*, ongoing project. Courtesy of the artists.

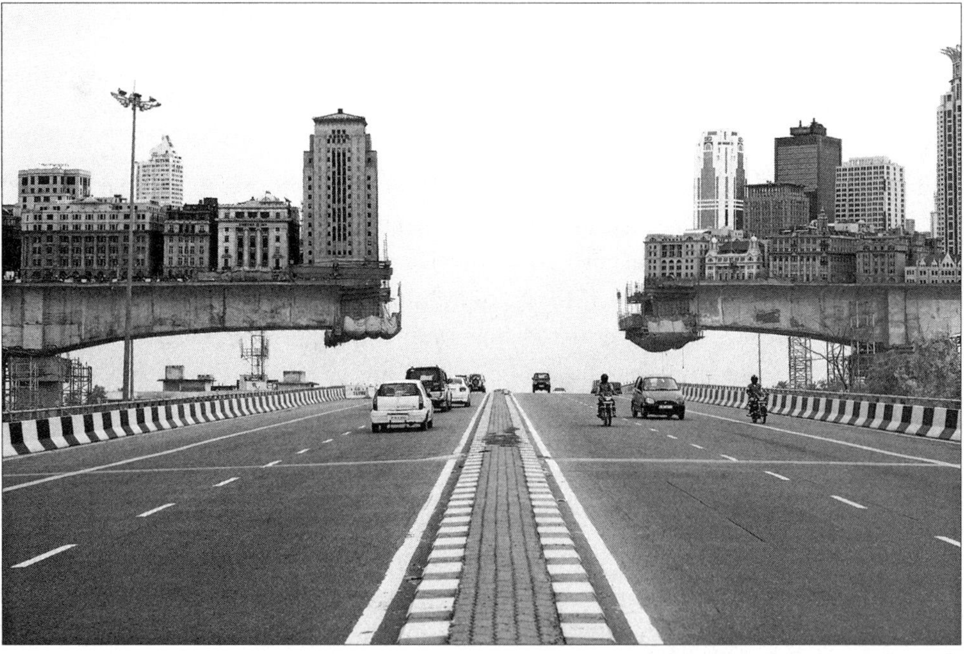

Figure 4.3
Gigi Scaria, *Face to Face*, 2010. Courtesy of the artist.

Scaria's work includes both images of actual mobile media use and urban landscapes rendered mobile. His images are patterns of both recognition and misrecognition, visualizations of the urban in decay. They are epitaphs to the cost of rampant consumption on the environment.

Despite a variety of new and remediated approaches by artists and collectives to comment on urban environments and the costs of high consumption, scholars have paid little attention to the ways in which artists have been using mobile media in critical projects. As we demonstrate in this chapter, mobile media are often now used as part of arts practice, in ways that go beyond their everyday uses for communication and content. They are deployed in ways that are sometimes related to, but often depart from, their uses in social activism. Indeed, in the Asia-Pacific arts context, mobile phones have become part of a complex entanglement between materiality, practice, and contestation that is emerging across the region.

Mobile Media and the Visualization of the Environment in the Asia-Pacific

The mobile phone is certainly implicated in the ways in which environmental issues and disasters are experienced across the region, yet this role plays out unevenly across different national, political, and economic contexts. In chapter 3, we noted that there are variations in how nation-states are involved in resource extraction, manufacturing, and waste disposal related to the supply of mobile phones for middle-class consumer cultures. These discrepancies in production and consumption are related to inequalities in the region, as they represent disparities not only in individuals' ability to consume but also in their ability to negotiate the price of their own labor. These inequalities are equally evident in the ways in which environmental disasters become connected to mobile media.

In some contexts, mobile media, their apps, and the social media platforms that connect users become active elements in the experience, management, and aftermath of environmental disasters. As mentioned above, in Japan, in the aftermath of the earthquake, tsunami, and subsequent Fukushima nuclear reactor disaster in March 2011, mobile phones were used extensively to capture and disseminate images around the globe. For many people, Twitter messages and still and moving images taken with camera phones not only embodied the effect and affect of the disaster but also became the repository for various forms of personal and communal grief and bereavement (Hjorth and Kim 2011). Known as 3/11, the event consolidated the highly symbolic and practical role of mobile media in everyday Japanese life. With established national broadcast media like NHK withholding important information about the Fukushima disaster under instruction from the Japanese government, this moment represented a major shift as millions of Japanese turned away from broadcast media in favor of mobile media such as Twitter, Instagram, and Line.

While camera phones are increasingly becoming the main vehicle and repository for recording events relating to climate change, it is important to remember that not all countries share Japan's decade and a half of mobile Internet access. In chapter 3, we introduced the global and regional inequalities that framed the impact of the super-storm Typhoon Yolanda in the Philippines in November 2013. This striking contrast not only exposes how differently the image of disaster can be visualized but also brings to the fore the unequal ways in which technologies are both experienced in, and affect, different Asia-Pacific populations. The path of the typhoon traced some of the poorest areas in the Philippines, where 40 percent of the population lived under the poverty line ($A1.25). These Filipinos had little choice but to live dangerously close to the western Pacific shoreline to subsist through fishing. Unlike the case of 3/11, in which

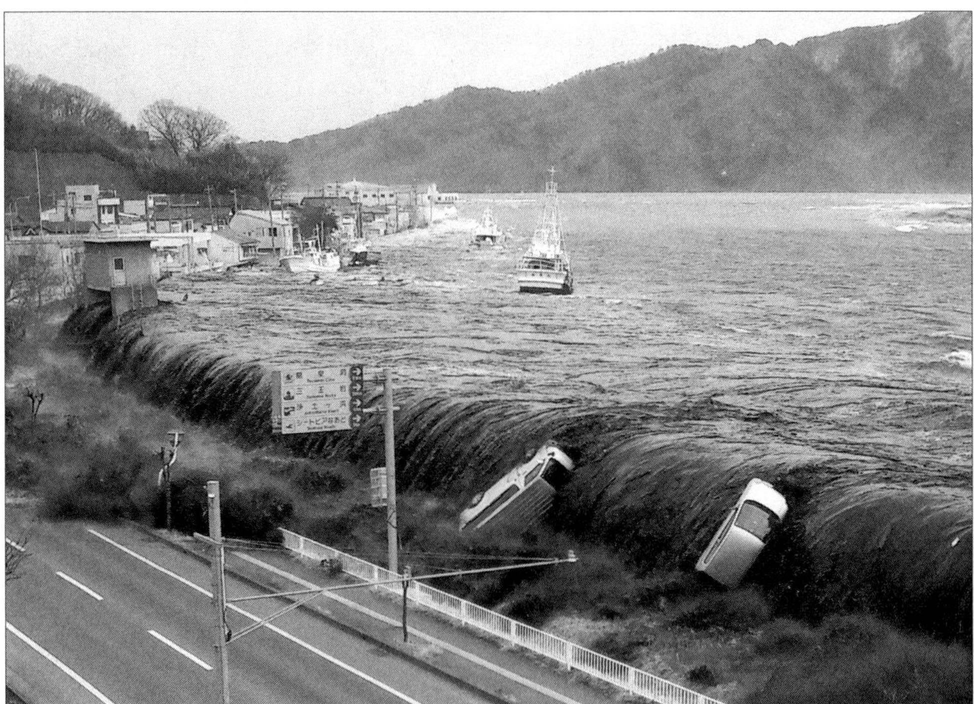

Figure 4.4
Mobile (*keitai*) media capture the horror of 3/11. Photo: Getty Images.

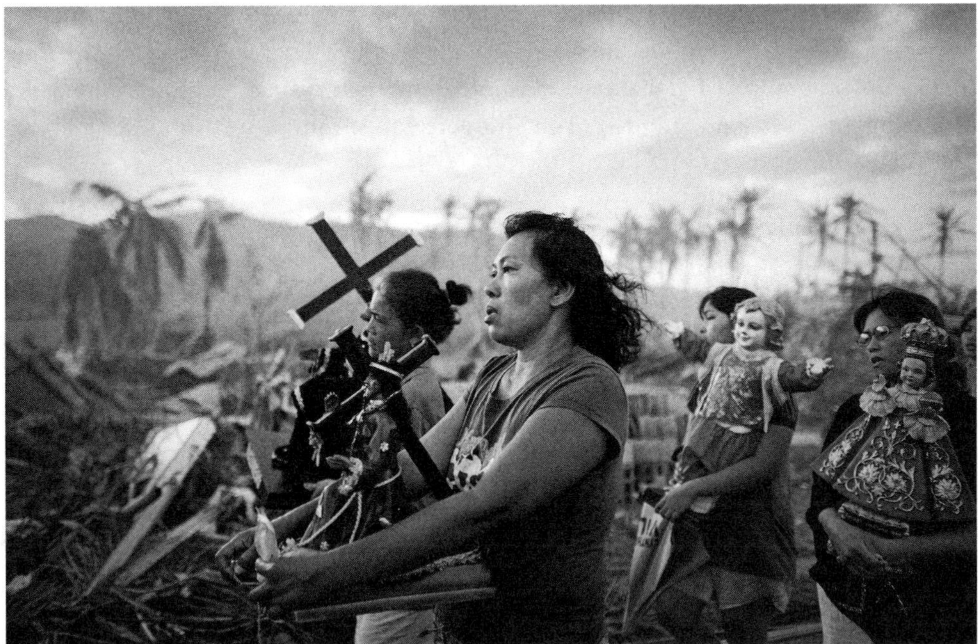

Figure 4.5
Images of Haiyan's aftermath. Photo: AFP. Photographer: Philippe Lopez.

images of the disaster were disseminated globally, in the Philippines an initial lack of images surrounding Haiyan was due to the fact that many affected people shared one mobile phone among a family, and most of these phones did not have cameras. Instead the images that emerged from the disaster were by professional photojournalists. They were well conceived in terms of the conventions of photojournalism, with attention to composition, and could be interpreted as being highly contrived. These were not the highly personal, intimate, DIY images associated with mobile media.

In chapter 1, we suggested that the smartphone can be understood as generating a sense of intimacy with the contexts, events, and persons to and with whom it connects users. In the context of environmental disasters such as 3/11, the smartphone as camera phone can be both a witness and an alibi to unfolding events. Smartphones have multiple apps, which enable the taking, editing, and sharing of images almost instantaneously. In comparing and contrasting Japan's 3/11 with Typhoon Haiyan, we can see that the uneven way in which mobile media have been disseminated across Asia also makes for the formation of different digital visual cultures and forms of social media engagement around disasters and their management. People in locations such as

Japan and South Korea have been high consumers and producers throughout much of the world's innovation around mobile media. Places like China have been the site for a significant proportion of global new technology manufacturing, and in locations like the Philippines, we are reminded of the immense spectrum of wealth existing across the Asia-Pacific.

Material and Immaterial Dimensions of Media and Art

We now turn to the question of how artists in the Asia-Pacific have engaged (with) media in relation to environmental issues to set the scene for a focus on how mobile media are being used. We are concerned with media not simply in relation to their content dissemination and communication functions but also as technologies with material and intangible affordances. This means attending to their implications as software and hardware, as well as to the ways in which the development of these technologies is framed by existing discourses, power configurations, and practices. Analyzing media from this perspective is also important for understanding how it might be implicated in processes of change. As Cubitt argues, being aware of "how software integrates with hardware, and how governance of standards impacts on engineering decisions," is essential for us to be able to comprehend how global norms are dominating the ways that we perceive the world. Once such an awareness is achieved, Cubitt suggests we would be in a position to "begin to remake those media in the interests of both artistic creation and the democratic and affective role of screen media in the twenty-first century" (2014, 134). The examples that we present in this chapter do not extend to the more radical aims that Cubitt expounds, but they do begin to demonstrate what kinds of critical and digital art projects have been enabled by existing media. They also signify how and where a remaking of screen media software and hardware combinations to suit an artistic and democratic and environmentally sustainable world might begin.

As we showed in chapters 1 and 2, across the Asia-Pacific, artists are deploying art to explore visualizations, representations, and understandings of our environment through a range of media and techniques. In this chapter, we focus on how artists are highlighting the connection between older technologies and newer media technologies in processes of visualizing the environment. This refashioning of media is a key element in generating a wider field of art practice that is emerging as a critique of technology, development, and waste in the region.

The Chinese artist Yao Lu uses technology as a medium through which to correlate environmental disaster with processes of development and modernization. Yao Lu's

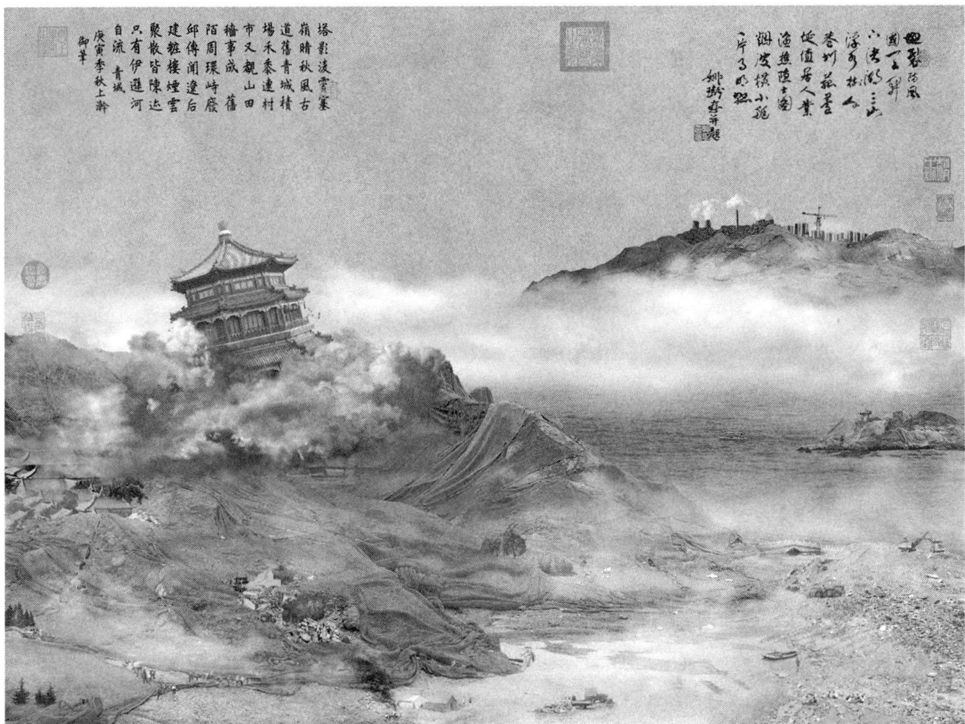

Figure 4.6
Yao Lu, *The Beauty of Kunming*, 2010. Courtesy of the artist.

photographs initially look like traditional paintings from the Chinese Song Dynasty. However, on closer inspection, a different understanding of the composition of the images appears, as we realize that they are of rivers teeming with rubbish, factories excreting toxins, and landscapes full of contamination. In presenting this critical perspective on China's rapid modernization, Yao Lu's work cleverly builds on the tension between narrative painting in China and its worrying environmental realities while simultaneously toying with an ongoing dialectic between the representational regimes and possibilities of photography and painting. From a distance, the pictures look like painted traditional natural landscapes, but up close they are photographs of man-made ecological disasters.

Yao Lu's work highlights the power of the economy over environment within China's rampant modernization. The sinologist Harro von Senger has argued that "Yao Lu's brilliant photographs reflect the hitherto neglected environmental 'material needs' of the Chinese people. His photographs can be interpreted as a constructive critique of the narrow, economy-centered policy of his country" (cited in Hattam 2011, n.p.). Yao Lu's work is also particularly interesting in that he links contemporary issues of human-made ecological disasters with the history of China. He thus reinforces a critique advanced by Giovanni Arrighi (2009) of Western depictions of the twenty-first century as the "Chinese century." Arrighi instead reminds us that capitalism can be understood as part of a five-hundred-year cycle in which China's dominant role globally has moved to and from the center several times.

In India, the work of Ravi Agarwal, an artist, activist, and writer, likewise focuses on questions concerning waste and technology. Agarwal is the founder of the leading Indian nongovernmental organization Toxics Link, which campaigns on waste and

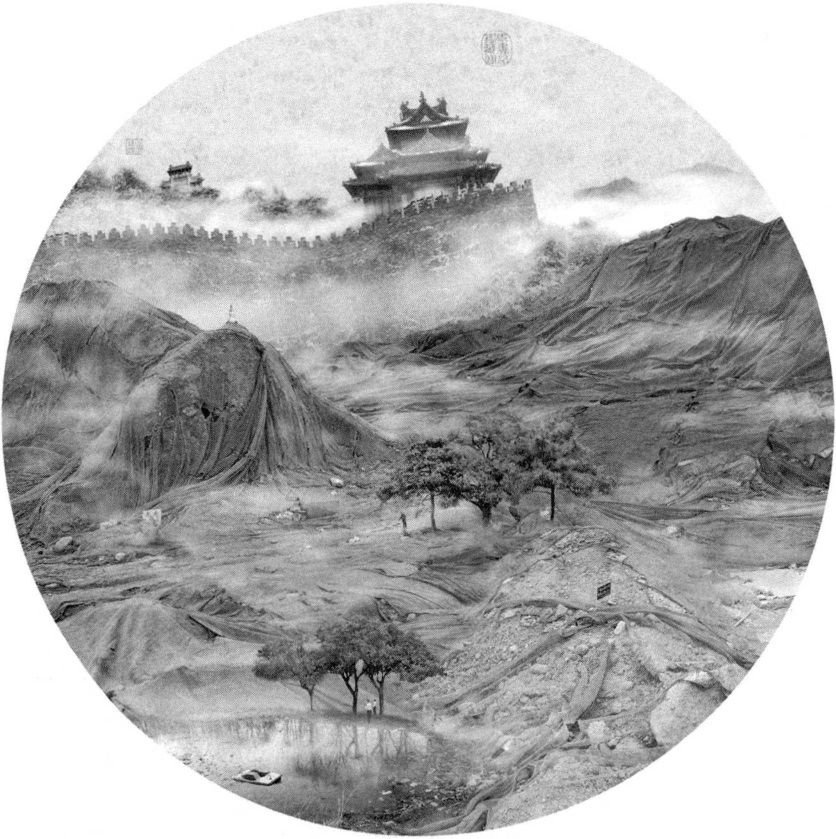

Figure 4.7
Yao Lu, *Spring in the City*, 2007. Courtesy of the artist.

ecology matters (http://toxicslink.org). In his photographic and video art, he uses social documentary to reflect on the tension between urban development and ecological sustainability. As Agarwal argues:

I am not suggesting that we go back to paradise, go back to nature. The challenge is what sustainability means. It's not only about the challenges of technology, of production or consumption, but challenges of self in relation to the environment around us. Social power and politics and ideas of self start influencing how we can imagine something, how we can change something. (cited in Brown 2014, 37)

For Toxics Link, e-waste is a key area of concern. As the organization's website notes:

India, currently, is estimated to generate more than 4 lakh tonnes of e-waste annually. The generation is estimated to go up many times in coming years, making it a critical issue. However, e-waste is not just a problem of waste quantity or volumes. The concern is compounded because of the presence of toxic materials like lead, mercury, cadmium, certain brominated flame retardants (BFRs) and many other chemicals. ... In a developing country like India, most e-wastes land up in the informal sector, where it is recycled without any consideration to health and

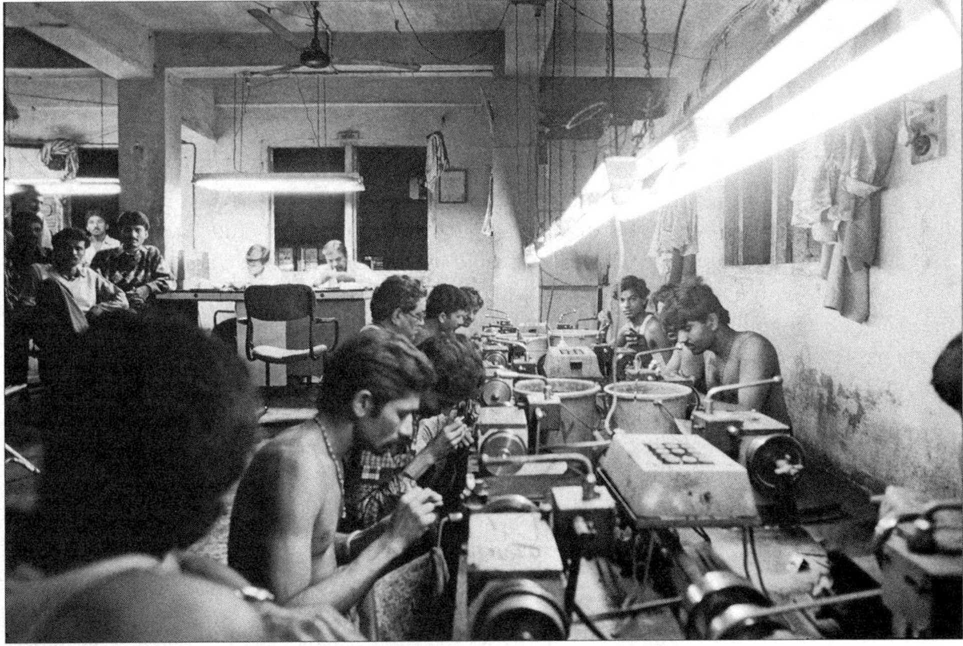

Figure 4.8
Ravi Agarwal, from the series *Down and Out: Labouring under Global Capitalism*, 1997–2000. Courtesy of the artist.

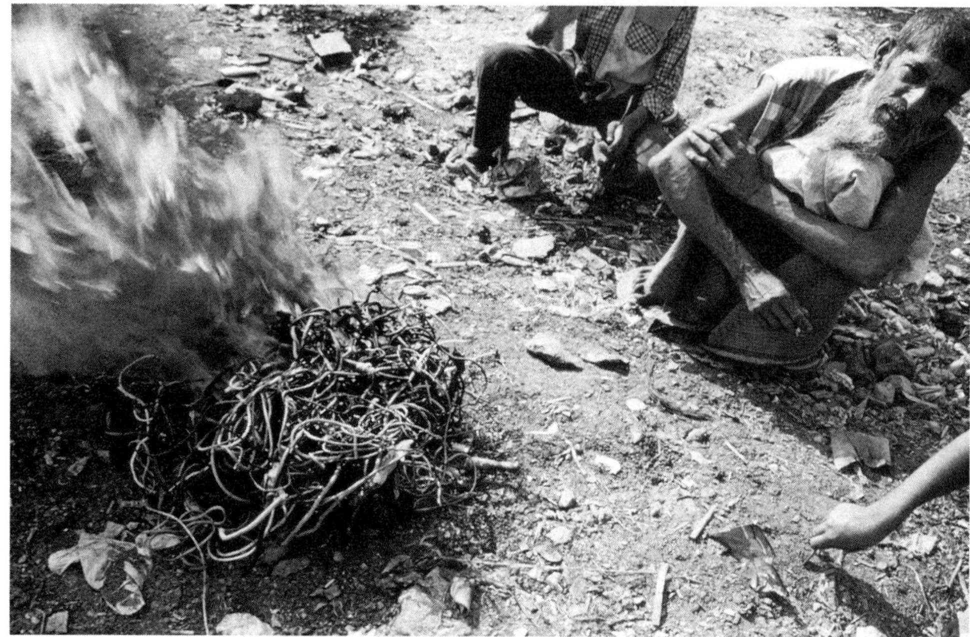

Figure 4.9
Ravi Agarwal, from the series *E-waste Recycling in Delhi* (Toxics Link, 2013). Courtesy of the artist.

environment. Open burning acid baths, unventilated work spaces and crude handling of chemicals are typical of these operations, where susceptible groups like children and women are regularly employed. With no safety equipment at hand, the workers in some of the recycling hotspots spread all over the country, are exposed to the toxic cocktails daily. The unregulated practices also release hazardous materials in air, water and soil, thereby endangering our environment. (Toxics Link 2014, n.p.)

Agarwal's practice is indicative of how some artists in the region are exploring the material and immaterial dimensions of the intersection between new and old media practices and their relationship to the environment. The work of the Japanese artist Takahiro Iwasaki also stands out in this regard. Iwasaki creates images of urban landscapes out of everyday objects. In one exhibition, he explored Kawasaki City (part of Tokyo Bay), which is known for its factories and for being part of the Japanese industrial backbone. Deploying everyday objects such as the bristles of toothbrushes, bath towels, and duct tape, he reconstructs intersections between the urban, the mundane, and industrialism.

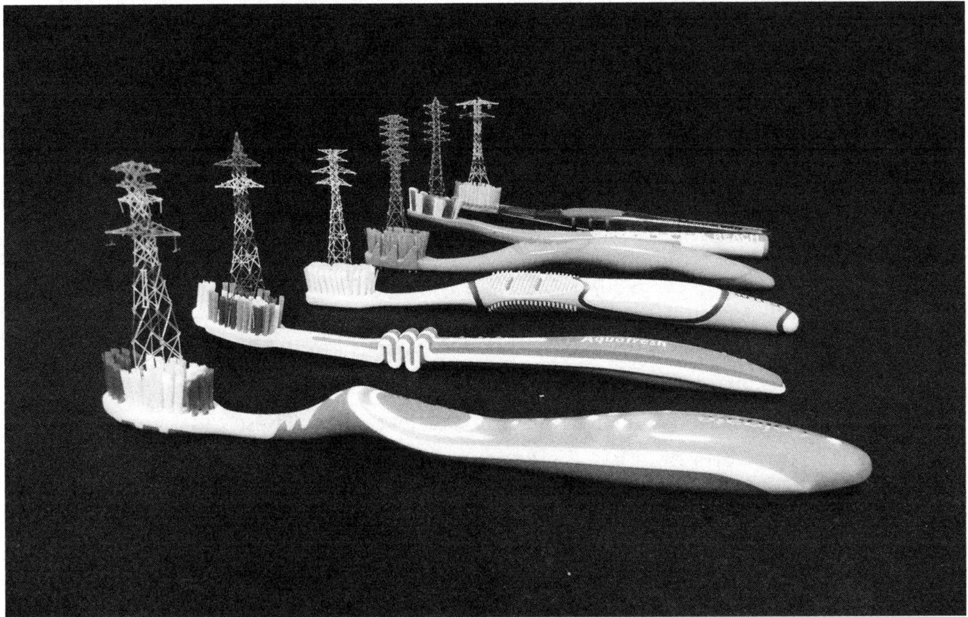

Figure 4.10
Takahiro Iwasaki, *Out of Disorder (bush)*, 2013. © 2013 by Takahiro Iwasaki, courtesy of Arataniurano.

Iwasaki's work marries the mundane with the sublime. He reappropriates and remediates the everyday objects he works with. His urban landscapes toy with scale to reinvent the environment and the everyday. Iwasaki draws on both physical sites and their online depictions, engaging with digital technologies both as media through which to produce his work and as the focus of his critical interventions.

The work of all the artists discussed in this section has inevitable connections to the very technologies and processes of technological and industrial development that they critique. Both the critical awareness and activist commitment to change invested in these projects, which aim to contest technological and industrial ills, work alongside and within constraints framed by the industries and political economies of which they are a part. In the next section, we explore how these relationships are manifested in a different context by examining the tensions generated when artists, media, and UCC become active participants in challenging norms of visualizations.

Artists, Media, and UCC

Although the adoption and use of mobile media remain uneven across the Asia-Pacific, they are increasingly being characterized by smartphones with online capacity (Hjorth and Arnold 2013). This is particularly so among the urban and increasingly affluent makers of, and audiences for, the forms of public art discussed in this book. For the artists whose work we discuss here, smartphones are increasingly established as part of a critical digital arts practice. For the people who experience and engage with their work—their publics—smartphones are already embedded in everyday life practices and relationships, making it an easy step to also use them in response to an artwork. However, there are two key tensions around the use of mobile media in relation to arts practice that have affected the way the field has developed. These tensions are derived from debates about arts practice. One such tension is between the artists who view media technologies as a vehicle only for dissemination, and those who actively use digital technologies, apps, and social media platforms as part of their practice as well as ways to engage (with) publics.

Another tension is demonstrated through current debates concerning the relationship between UCC, amateur creativity, and professional practice, which have also slowed the uptake of mobile media among some artists. Some art critics, such as Claire Bishop (2012a), have neglected to account for how new media have become an integral part of art practice. This sort of move exemplifies how definitions of art in relation to new media are still being contested. As Hjorth's (2013) study of the role of social and mobile media in the work of artists has also revealed, practicing artists and creative UCC producers tend to have different understandings and uses of social media. Much of the work around UCC has focused on amateur creativity (Burgess 2007; Burgess and Green 2009), thus reinforcing amateur and professional divides. New media artists have been quick to adopt social media platforms such as YouTube and Facebook both as a *frame* for context and content and as a *medium*. In contrast, according to Bishop (2012a), visual art practitioners have been less enthusiastic. For them, YouTube is often simply a site for distribution and dissemination, not a medium with its own context and content for generating new forms of art and visual culture.

In this context, however, some groups of visual artists are increasingly using mobile and social media apps and platforms as vehicles for, and of, popular culture. For example, a new global breed of artists who call themselves "mobile masters" and "social media artists" (such as Man Bartlett, New York based but "ubiquitous" online via Twitter) are using mobile media in their work in ways that are analogous to working

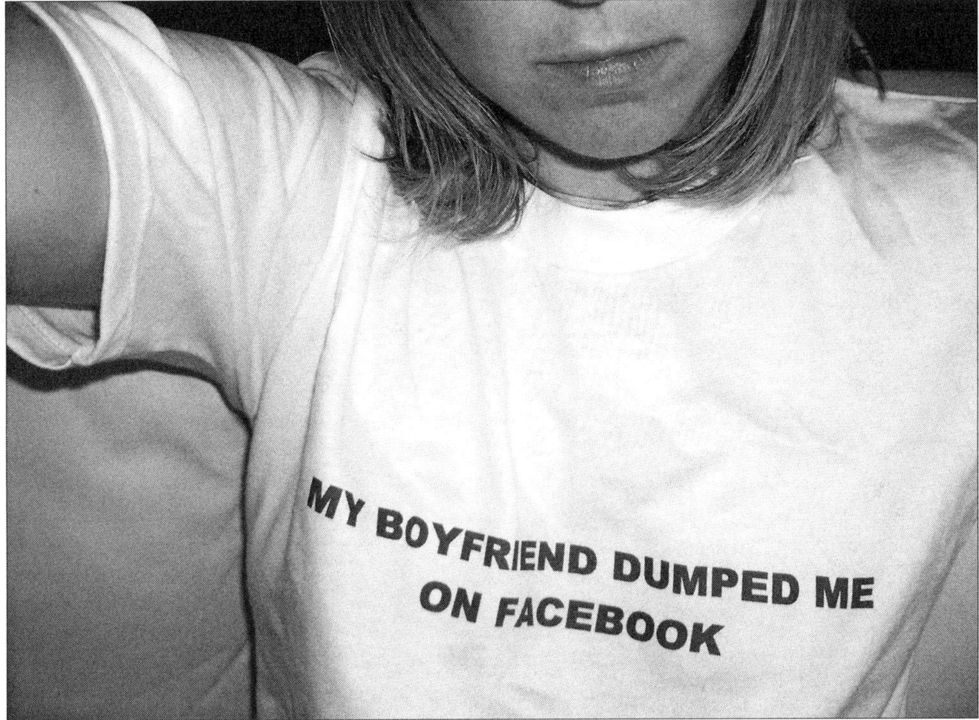

Figure 4.11
Anastasia Klose, *My Boyfriend Dumped Me on Facebook*, 2007. Courtesy of the artist and Tolarno Galleries.

with a canvas or in a photography lab. For the Melbourne artist Anastasia Klose, social media such as Facebook are part of a broader "intimate public" turn (Hjorth, King, and Kataoka 2014). Drawing inspiration from feminist performance artists, Klose's work engages in and through social media as part of teasing around contemporary popular cultures. Like the UK artist Tracey Emin, Klose blurs the personal and political, often using her search for love to engage audiences across various media.

The work of the New Zealand artist Janine Randerson is indicative of how a generation of new media artists are using Big Data visualization to expose the realities of climate change. In 2008, Randerson undertook a residency at the National Environmental Research Institute in Roskilde, Denmark, where she monitored the effects of climate change on animal and bird migration. In her video installation *Cascade*, Randerson brought together the two disparate worlds of scientific data visualization by scientists with found amateur UCC footage from YouTube. She thus combined in her work the amateur and the professional, and the objective and the subjective. In doing so, she denaturalized the persuasiveness of Big Data visualizations.

Like Randerson, the Australian artist Jeremijenko explores the entanglements between new media and the environment. As one of the early participants in net.art (an important site for early new media online art), Jeremijenko has always been at the edge of technological innovation, design, and social commentary. Self-described as an "X designer" (experimental designer) and "thingker," she deals with everything from genetic engineering to biodiversity (Brown 2014). As Jonah Weiner notes in his *New York Times* piece on Jeremijenko:

Much of Jeremijenko's work reimagines environmentalism as a kind of open-ended game. She likes to frame ecological arguments not in the common conservationist language of interdiction—consume less, reduce your food miles, lose those incandescent bulbs—but rather as overtures to more engaged and imaginative participation. "A lot of my work concerns a crisis of agency—what can we do?" she says. Her art, which has appeared at the Museum of Modern Art, two Whitney Biennials, the Guggenheim and her New York gallery, Postmasters, can take the shape of fanciful provocations, prototypes for functioning systems or, as often as not, both. The idea, she says, is to design interfaces or write scripts that will "facilitate interactions between humans and nonhumans." (Weiner 2013, n.p.)

Jeremijenko developed such a form of play between human and nonhumans in her work *Mussel Choir*, mentioned in chapter 1, which was installed in the East River in New York and Melbourne's Docklands in Australia. This work is particularly interesting as an example of UCC in that nonhuman actors become part of the process through which the content of the artwork is generated. The work contains sensors that, depending on whether the mussels are open or closed, automatically play a particular sound track. Mussels respond to the water quality, and the better the water quality, the more they open and sing, whereas when the water quality is lower, the mussels close and hum. Jeremijenko's work highlights the way in which media art can explore the multidimensionality of the environment. Her scenarios of use combine the scientific with the artistic to probe and question. This probing of key questions around climate change is highlighted in the Melbourne-based arts project Climarte (2015). Clearly located within advocacy, Climarte "harnesses the creative power of the Arts to inform, engage and inspire action on climate change" (Climarte 2015, n.p.).

As the examples discussed in this section show, far from only serving as a vehicle for dissemination, digital media and platforms are increasingly becoming part of the ways in which artists connect with both the world and the people and nonhumans in it. These forms of arts practice bring the agency and activity of publics into spaces traditionally reserved only for artist-produced works. By using digital technologies and

A   B

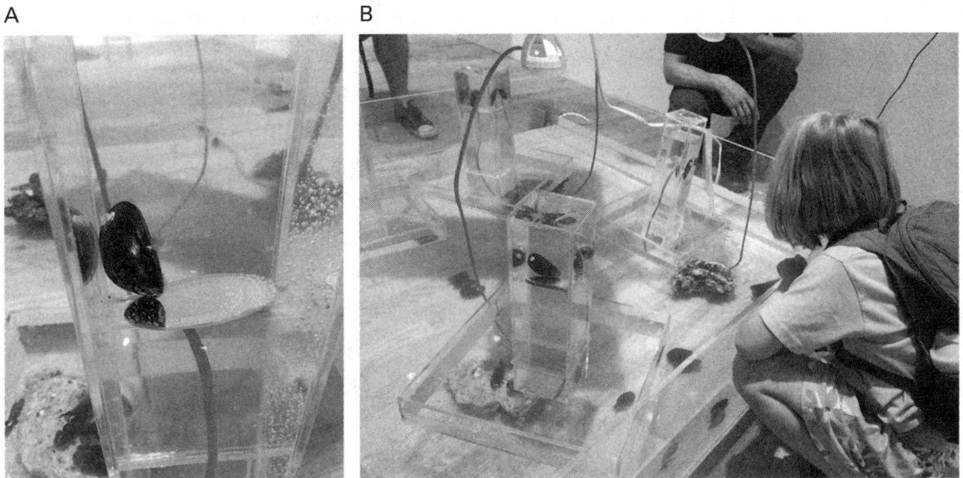

Figure 4.12
Natalie Jeremijenko, *Mussel Choir*, 2014. Courtesy of the artist.

media as conduits through which to introduce UCC into art spaces, these Asia-Pacific artists are shifting the ways in which art can serve as a critical force in the context of climate change and environmental disaster. In doing so, they are also inviting new forms of public engagement with, closeness to, and participation in, these issues and the development of critiques around them.

Camera Phones: The Artist, the Amateur, and the Everyday User

One of the key modes through which amateur creativity and professional creativity are being reconciled is in the use of the camera phone app. However, apps need not only to be understood as technologies that enable better or more participation, connection, and communication, but also to be interrogated critically. In this section, we discuss how apps are implicated in the development of social relations and cultures of participation, but first we frame these developments within a consciousness of how apps themselves are situated within debates about the environmental consequences of media and development.

Apps are increasingly being used to connect artists and amateur image producers and, in doing so, to create critical artwork and commentary. This brings to the fore an issue that we highlighted at the beginning of the chapter; that is, to understand how media practices are framed ideologically, we need to attend not simply to the technological hardware but also to the software that makes such connections possible. This is necessary for us to interpret the political and economic contingencies through which they are actually used, or which they might be appropriated to resist. Camera phone apps are now ubiquitous in the everyday lives of most smartphone users, providing ways of sharing, archiving, and participating across a range of social media platforms, as well as in more private digital domains.

Photo-sharing apps are rapidly becoming key players in this field. For example, the photo-sharing and social networking app Instagram has quickly emerged as a player in some social media ecologies of the Asia-Pacific region. Of the 200 million active Instagram users, 80 million are located in Asia (Instagram 2013). In Thailand, where in 2014 there were 24 million Facebook users, the number of Instagram users increased from 240,000 in 2012 to over 1.5 million in 2013. Moreover, Instagram is not only used by locals. Tourists have been active users, making Bangkok, the capital of Thailand, one of the most photographed locations in 2013 and the second most Instagramed city in the world (Instagram 2013).

Camera phone apps and the images they produce are used for documenting, sharing, communicating around, and critically commenting across a wide range of contexts, relationships, and issues. These range from environmental disasters (as discussed at the beginning of the chapter) to everyday experiences of the environment, as we discuss hereafter. Yet just like smartphones themselves, apps are part of, and contribute to, the perpetuation of the consumer society that is a contributor to climate change. It is due to apps like Instagram that an astounding quantity of digital data is produced and needs to be stored every year.

Apps are therefore positioned in a particular way in the media ecology, which, as Maxwell and Miller (2012) have pointed out, is part of the problem constituting climate change and environmental crisis. This is one of two key points we need to keep in mind when considering how camera phone apps might also be used to critically engage with climate change. While this issue has not yet been confronted by scholars in the field of mobile media and camera phone studies, the second point brings us to a much more established area of research. Camera phone apps and images can potentially bridge the gap between amateur and artist images to make critical interventions that involve publics in debating and contesting climate change issues. Existing research in mobile media studies of everyday users provides a starting point for considering these possibilities.

In their earlier studies of camera phones, Mizuko Ito and Daisuke Okabe (2005) argued that what they called the three *s*'s—sharing, storing, and saving—played a pivotal role in the way these technologies were used. These, in turn, informed the production of what was predominantly "banal" everyday content (Koskinen 2007). As camera phones have become more commonplace, along with the explosion in smartphone ownership and use, new contexts for image distribution, such as microblogging and LBS (location-based services) have been established. Subsequently, new types of visuality have emerged. In this section, we discuss two of these: intimate visualities and emplaced visualities. The concept of intimate visualities refers to forms of digital photography that bring localities and experiences together with users' personal journeys and forms of self-expression. This concept is significant for understanding how everyday users might participate in public art projects through camera phone images, because it brings our attention to the ways that camera phone images engender forms of contact and closeness at a personal level.

These new visualities situate users at the center, and in doing so, they produce personalized and intimate ways of mapping out the world as photographed. These are often shared online through the platforms that apps are associated with. The concept of "emplaced visualities" is complementary in that it focuses similarly on the relationship between the photographer and her or his environment, but it does so in a way that attends more closely to how these environments are constituted and how users fit into these contexts than to how sentiments such as intimacy are generated through them.

The forms of intimacy generated through the use of camera phone apps and geotagging are often social, as images are uploaded to social and photo-sharing platforms such as Instagram and Flickr or other social media platforms such as Facebook. When shared on social media platforms, camera phone images can generate forms of being together online among users, which in the mobile media studies literature is referred to as digital copresence. This is a way of sharing presence online with a person or persons while not being together physically in a material locality. Rather than privileging face-to-face contact, the concept of copresence focuses on the mediated nature of all interactions.

Scholars in this field, such as Christian Licoppe (2004), have situated copresence as part of a process of the evolution of mediated intimacies. In a similar vein, Jason Farman has suggested that mobile technologies have transformed our experience of presence and absence into "a social proprioception" of perpetual copresence (2012, 108). With LBS added to the digital mapping process, physical presence is also acknowledged. This has led researchers in this field to emphasize how corporeality is brought to the fore, for instance, as "a social network that engages users as embodied interactors rather than disembodied voyeurs" (Farman 2010, 869), or in our own work, as constituting "emplaced visuality" (Pink 2011).

Another way in which camera phone apps and digital platforms are generating new forms of closeness and senses of proximity between users is in the way that they question, challenge, and are capable of breaking down divisions between amateur and professional paradigms (Hjorth 2007, 2009). They suggest a democratization of media, often offering cheap and convenient alternatives for those who cannot afford digital cameras. For example, in Seoul, women who were previously not interested in photography are developing a passion that aspires toward the professional (Lee 2005). So, too, within online community spaces like the digital photography sharing platform Flickr, strangers and intimates are sharing images and comments (both aesthetic and technical) in ways that suggest that new visualities are ordered by a "situated creativity" (Burgess 2007) and banality (Mørk Petersen 2008) generated through that very context. As Edgar Gómez Cruz and Elisenda Ardèvol's (2013) discussion of an ethnography of Flickr photographers has also shown, this platform also creates a democratization of professional and amateur status, whereby boundaries between the two can become blurred.

While social media platforms develop other ways of allocating status and hierarchy, they have also been used to facilitate new forms of communication and closeness between persons of different status and standing, which would be impossible in most face-to-face contexts. For example, we have already noted how the Chinese artist Ai Weiwei has used Twitter to create critical debate internationally. Through the Twitter platform, he has been able to communicate directly with his public beyond the mediations of art institutions, government institutions, and dealers. While such examples represent exceptions rather than the rule within the professional art world, they reveal the potential of social media to bring together these often disparate spheres, and as we show in the next section, in the work of mobile media artists, the relationship between artist and public is further broken down.

In chapter 2 we outlined how a theory of place that accounts for the digital elements of the environments we inhabit can bring together the experiential and representational within a single framework, which in turn relates to concepts of space, scale, time, movement, and perception. By seeing people and technologies as equally emplaced within such environments, we can start to consider how forms of closeness and intimacy generated through the use of camera phones and social media are part of wider ecologies of place, and thus to conceptualize camera phone photos as a form of *emplaced visuality*. For instance, when we use camera phone apps like Instagram, where geotagging almost becomes the default, apps begin to play an active role in how places are conceptualized, mapped, and experienced. Through camera phone geotagging practices, the temporal and spatial dimensions of places become entangled with elements of experience associated with digital and social media, and make explicit how the place that the user and image share is one that traverses the digital and material-physical architectures and dimensions of the environments we inhabit.

Before exploring how camera phone users have been engaged in producing images as part of an art project, we outline our understanding of how ordinary uses of camera phones are developing in everyday contexts. This background, we suggest, provides a series of insights that enable us to further our understanding of the kinds of existing practices that will subsequently inform the ways in which everyday camera phone users are likely to participate in public art projects. Through a series of camera phone ethnographies undertaken online in Australia, Hjorth and Pink (2014) have explored how camera phone apps are used within everyday life, focusing particularly on how participants used their camera phones to take and upload images as they move through the world.

Drawing on this work, we argue that camera phone photography is (at least very often) an outcome of how people move through their everyday environments and the intersections they make with other persons and things in movement. Such photography takes place *in the world* rather than capturing images *of the world*. As Hjorth and Pink (2014) have shown, it emerges from the relationships between people and their environments. The default geotagging of camera phone images locates them within digital maps, which are themselves representations of the world. This means that such images are also, in an explicit and direct way, not only *of* the physical world but also situated *in*, and part of, the digital and representational environment that is merged with what we would conventionally have thought of as a "natural" environment. To explore how these forms of materially and digitally emplaced visuality are produced, Hjorth and Pink have developed the concept of the *digital wayfarer*.

For the anthropologist Tim Ingold, "The path of the wayfarer wends hither and thither, and may even pause here and there before moving on. But it has no beginning or end. While on the trail the wayfarer is always somewhere, yet every 'somewhere' is on the way to somewhere else. The inhabited world is a reticulate meshwork of such trails which is continually being woven as life goes on along them" (2007, 84). Developing Ingold's notion of wayfaring, Hjorth and Pink propose that such forms of movement can equally be used to comprehend how, in a digital-material environment, everyday camera phone users repeatedly traverse the online and offline. This concept puts a theory of movement at the center of our understanding of contemporary media practice and accounts for the idea that people take, and often post, camera phone images when they are moving through the environments that they live in. The camera phone images exist between the online and offline in a place that is constituted through the relationship between them.

As we noted earlier, however, being online with locative media is often associated with forms of sociality, which scholars have referred to as digital presence and copresence. If we likewise consider wayfaring, with its exploratory stance, to have an element of sociality, and thus understand moving through the world as inherently social as well as embodied and emplaced, we can make a theoretical connection between the visuality and sociality of the intersection between camera phone photography and locative media practices. As we will see in the following sections, these possibilities not only play out in everyday life but also have potential in the making of mobile art.

Finally, because everyday users have begun to participate in producing online understandings and representations of place, we see a shift in how cartography is being defined. As we must do for all software, we need to account for how digital mapping is, on the one hand, a corporate and institutionally driven and regulated field (e.g., Lapenta 2011; Pink 2011). Critical approaches to cartography both challenge the authority of traditional mapping and interrogate new and digital mapping practices to investigate the institutional and power relations that frame them. Yet there is also a powerful argument for seeing uses of geotagged camera phone apps like Instagram and Snapchat as producing contested modalities of presence that disrupt linear readings of maps and instead highlight the collaborative, performative, and creative dimensions of cartography within mobile media (Lammes and Wilmott 2013; Verhoeff 2013; Wilmott 2012). With shifts in technologies and the rise of collaborative platforms, the activity of making and sharing maps has taken on new playful, ambient, and copresent dimensions. Camera phone apps are part of what Kitchin, Dodge, and Perkins (2011) have called cartography's "transition" tool kit, used by everyday cartographers who are sharing, remixing, and adding to our understandings of media geography.

As we have outlined in this section, the social, spatial, and critical potentials of camera phone photography are being played out across a range of everyday life contexts that, as a matter of course, traverse the online–offline or digital-material environment that we inhabit. This is no less so in the Asia-Pacific than elsewhere. In the next section, we reflect on how these implications of camera phone use are also being harnessed for critical and participatory arts practice in the region.

Art, Play, Mobile Media, and Intimacy: *Shibuya: Underground Streams*

In the cultural studies literature, mobile media have been associated closely with the generation and maintenance of new forms of intimacy and ways of being together online, called digital copresence. The capacity of mobile media to generate forms of intimacy is also being developed through the work of artists who engage with these technologies. In a series of online interviews, Hjorth identified how artists were developing templates through which forms of creativity and intimacy were being entangled in new ways. From intimate media such as camera phones (Pound) to familiar tactics like Web 1.0 flash repetition (Candy Factory) and popular cultural mimicry (Cao Fei and Klose [see fig. 4.11]), new technologies are creating new forms of intimacy (emotional, cultural, psychological proximity) in both the production and use of mobile phones in these art projects (Hjorth, King, and Kataoka 2014). Cao Fei's work deploys a creolizing of the online (Second Life) with the offline by playing with stereotypes around East (i.e., communism and pandas) meets West (rapid consumption). Here commentary on transnational consumer narratives has, more recently, focused on a tension between the environment and media characters. In figure 4.13, characters from the children's program *The Night Garden* are homeless, with rubbish as their furniture. Unlike in the TV program, in which the environment is beautiful, Cao Fei's characters have met with a cold, hard reality.

This propensity toward intimacy in mobile phone art has also been created through the game-like qualities of some mobile media art, such as the *Shibuya: Underground Streams* project, discussed hereafter. As Chris Chesher (2004) has noted, mobile games create new forms of engagement centered on a type of intimacy in which the visual is no longer defined through twentieth-century paradigms such as the gaze (film) or the glance (TV) but rather through the *glaze*, a visuality that also involves other intimate and "sticky" practices such as the haptic (touch) (Hjorth 2008b). Notwithstanding the wider possibilities for theorizing the sensory and emotional affordances of mobile media and smartphones (which is not our focus here), the implications are that these technologies can therefore be thought of as offering a new form of embodied and sensory engagement with art, and this needs to be understood beyond the idea of users simply providing digital visual content to a digital visual arts project.

Instead, we must account for these practices as a far more sensory, emotional, and *intimate* engagement or form of participation. We now demonstrate how these uses of mobile media in art can be played out in relationship to actual events and environmental issues by returning to the story of the 3/11 earthquake and tsunami in Tokyo. We continue this account to show how social and visual media, smartphones, and arts

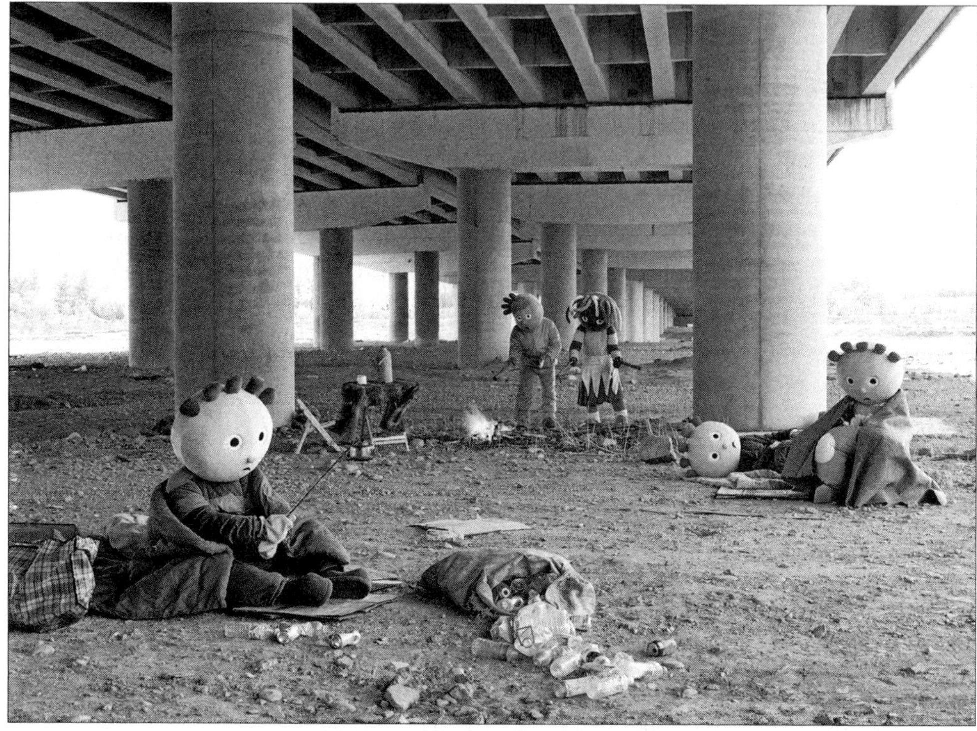

Figure 4.13
Cao Fei, *After a Long Day*, from the *Post Garden, Eye-SPY* photo series, 2011. C-print. Courtesy of the artist and Vitamin Creative Space.

practice have become entangled in the aftermath of 3/11. Our starting point is the account of "Toshi," a participant in Hjorth's research in Tokyo who was playing a haptic game during 3/11:

When the earthquake occurred, I was alone in my room, playing a monster hunter PSP game. Exactly at the time, I was fighting with a monster who makes an earthquake, so that I didn't realize that an actual, offline quake had occurred. Only after beating down the monster, I realized something different around me. A fish tank had overflowed, and books had fallen down. Initially I was not really shocked by the earthquake itself but felt frustration with the aftermath—the power failure, panic buying, nuclear accident, and such stuff. During this time, I stayed inside with a friend and continued to play the monster hunter game. But the game was no longer entertaining. ("Toshi," twenty-five years old, Japanese male)

Toshi's immersion in the PSP game was so deep that he mistook the quake vibrations for the monster's movements within the game. In the moments afterward, he realized the horror of the real life event and desperately tried to contact friends and family, but to no avail. In the days after 3/11, as multiple and conflicting news reports emerged across mass and social media, Toshi and a friend used the game to hide from the pain and confusion.

Later it emerged that the national broadcaster, NHK, had deliberately withheld important information about the Fukushima reactor under the instructions of the government. For Toshi—like millions of other Japanese—his trust shifted toward mobile media such as Twitter, and LBS like Foursquare and Instagram, enabling him to feel closer to others who were also experiencing this aftermath, as well as to have a sense of continuous copresence with his family and friends. Toshi's gameplay at this time was geared toward intentional escapism, given that the world was too traumatic and confusing. It was also part of a wider configuration of textures of grief that swept Tokyo and its population after 3/11.

At this time, the Australian-based project *Spatial Dialogues: Public Art and Climate Change* began to collaborate with Japanese artists and activists through the BOAT PEOPLE Association (BPA). Through this collaboration, a project was conceived in Tokyo in a post-3/11 context. Titled *Shibuya: Underground Streams*, it sought to explore the role of

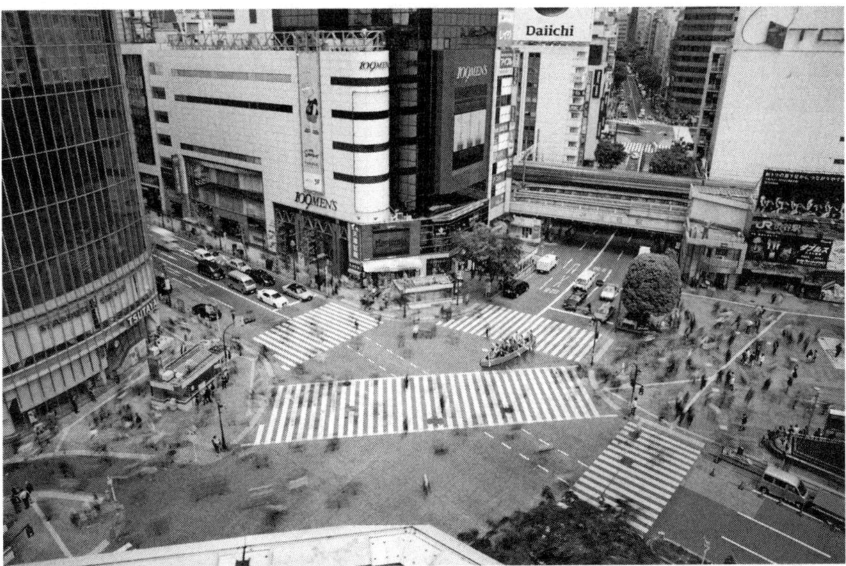

Figure 4.14

Shibuya: Underground Streams. This image shows a boat in one of the world's busiest intersections, Shibuya crossing.

art and screen media in perceptions of climate change in Melbourne, Tokyo, and Shanghai. The Tokyo 2013 collaboration involved Japan's BPA as well as individual Japanese and Australian artists. As part of the project, the mobile art game *keitai mizu*, organized by Hjorth, was informed by the question "How could we harness Twitter and camera phone apps to make a game that reflects on the environment in new ways in a Tokyo post-3/11 context?"

Japanese and Australian artists were asked to make a series of abstract and representational works of water creatures, which were then placed around a specific park in Shibuya. The project sought to disrupt dichotomies between art and nonart, water and nonwater, game and nongame, player and ethnographer. Players had fifteen minutes to hunt for various native-only, water-related creatures and objects that had been placed around the site. They then "captured" the art with their camera phones and shared it online on Twitter or Instagram. Winners only sent pictures of the native species to the *keitai mizu* Twitter account.

The *keitai mizu* project deployed two concepts that Hjorth has developed in her mobile media art—*intimate copresence* and *ambient play*—to engage with and reappropriate camera phone taking and sharing of images within Japanese everyday practices. Earlier, we examined how mobile media can enable new forms of intimacy through digital copresence. For the *keitai mizu* project, the concept of copresence is particularly useful because it deliberately conceives of presence as a spectrum of engagement across

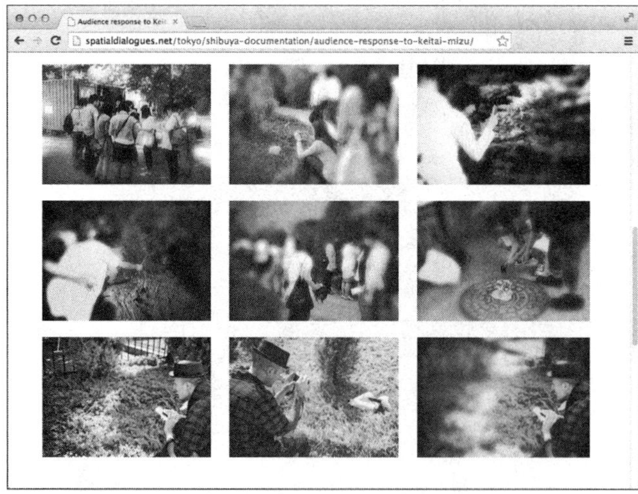

Figure 4.15
Keitai mizu players, 2013.

multiple pathways of connection. It thus goes beyond counterproductive dichotomous models of online and offline, here and there, virtual and actual. The concept of ambience is often used to describe sound and music, but it has also been used in computing and science to describe ubiquitous media (Dourish 2005). As a noun, it specifically refers to a style of music with electronic textures and no consistent beat that is used to create a mood or feeling. More generally, the term describes the diffuse atmosphere of a place, which can be thought of as an affective texture of place. In short, ambience is about the texture of context, emotion, and affect.

Ambient play therefore involves a blending of forefront and background, interwoven into the rhythms of everyday life. As Miguel Sicart (2014) notes, play and playfulness are increasingly becoming important within every facet of contemporary life. By playing with ambience, the game brought the park's hidden water streams to the forefront through the artworks placed around the park. Many of the participants were unaware of the rivers running beneath the roads and train tracks. The fact that, in many cases, it was hard to distinguish between the art and rubbish in the park meant that players were constantly having to analyze each everyday object as a potential artwork. This confusion meant that players often ended up talking to one another and asking other players' advice, as well as that of friends who were electronically copresent on Instagram and Twitter. This is a vivid demonstration of Kester's dialogical aesthetics. The slippages between the intentional objects of the game and what was already present in the park further defused the relationship between the art and nonart, game and nongame space. The artwork was ambient, moving in and out of the background while setting the mood.

Copresence was a key feature of the game. As players went around photographing and sharing images online, copresent friends (i.e., friends geographically absent but electronically present) began to collaborate and query the pictures taken. *Keitai mizu* thus sought to highlight the importance of ambient copresence in the construction and experience of place. Some digitally copresent friends were co-opted into the game space by trying to help their friends solve the game by guessing what was the art and which were the native Tokyo water creatures. The game space intentionally blurred online and off-line spaces with Instagram and Twitter, enabling digitally copresent friends to share the experiences and images. Through the process of gameplay, participants became more mindful of the local water species, as well as aware that the city is made up of numerous little rivers running beneath its train lines and roads. The artists wanted to make audiences aware of the hidden water cartographies and how mobile media apps demonstrate how maps are performative by suggesting the idea that we shape maps as they shape us.

Keitai mizu attempted to challenge boundaries between official and unofficial game spaces by blurring them with different modes of play (fig. 4.16). In particular, camera phone practices partake in new haptic visualities that bring emotional and social dimensions of place and play to the official gameplay space and drive the motivation for use. By deploying camera phone practices as part of the mobile game, players can develop the melodramatic elements—the affective and emotional dimensions—to engage friends in the play of being mobile.

Through the playful use of geotagging on Instagram, whereby numerous images of artifacts were assembled on the website, players were able to see other players' choices (i.e., what they thought were the native animals) and their location. This created a sense of emplacement but also displacement as other players searched for some art objects that were either mistaken for rubbish in the park or too small to see (some artworks, such as those by Yasuko Toyoshima, were semitransparent creatures that were

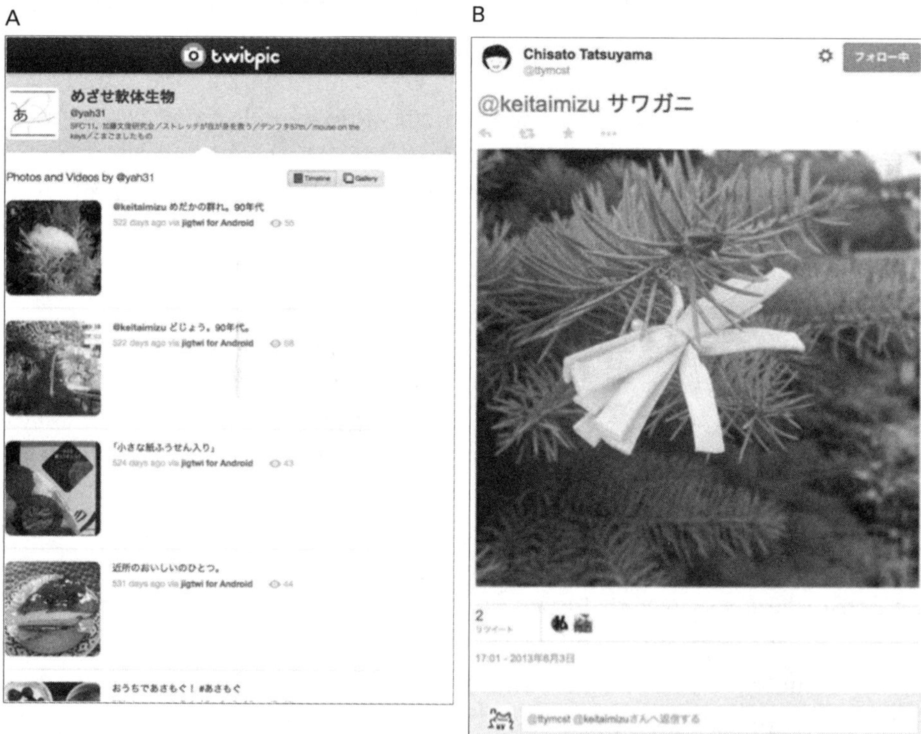

Figure 4.16
Copresent and ambient play examples of *keitai mizu*, 2013.

only five centimeters long). The *Spatial Dialogues* website can therefore be thought of as becoming part of the emplaced visualities of the park through each of the players' interpretations. The mapping of the park and its underground streams became a series of Instagram clues. These entanglements of ambient play, copresence, and emplaced visuality were essential components of the game. By deploying the unofficial but significant role of camera phone sharing within copresent spaces, this project demonstrates one of the ways in which alternative cartographies of urban environments might be constituted, in this case in the aftermath of an environmental disaster.

How people engage in a mobile gaming projects like *keitai mizu* through the use of camera phones becomes a representation of place but also simultaneously articulates the world in movement. That is, images are taken "on the move" and engage with embodied and affective relations of place, in this case the Shibuya River, but they also form part of the larger, imagined transnationalism discussed in the previous chapter. Platforms like Instagram offer a particular form of intimate public that can be deployed to reconceptualize the relationship between art, media, and the environment. In another mobile game, *mobile play* (fig. 4.17), which was part of the final

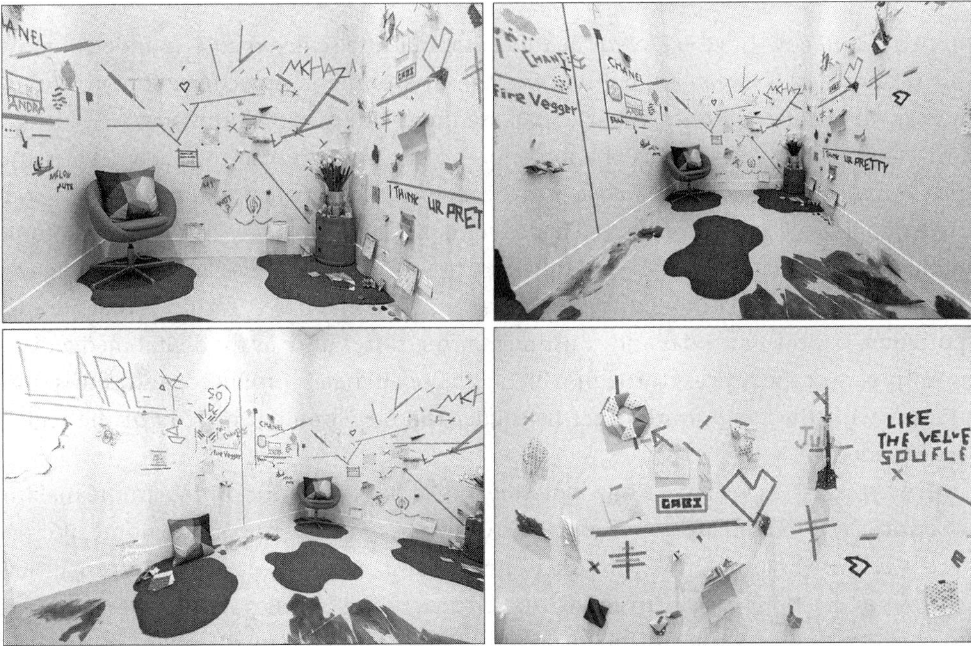

Figure 4.17
Playbour Inc., *mobile play*, 2014. Photo: Larissa Hjorth.

Spatial Dialogues project, water "stains" were placed in and around a gallery space in Melbourne. Players had five minutes to find and photograph all the water drops as they engaged with a Twitter bot mobile game program. Those who did not have smartphones could draw a map of where the water drops were on the wall with pens, paper, or sticky tape. In Australia, searching for water takes on a particular meaning, given the country's ongoing drought issues.

As these mobile play and art projects highlight, copresence articulates the politics of place as a platform. We take up the complex and often contradictory nature of place platforms, as a way to understand the entangled relations of media, art, and contemporary culture, further in chapter 5. To end our discussion of *keitai mizu* here, however, we emphasize that because it was played out in part on a public online digital platform, the project also constitutes an intervention that goes beyond its own subject matter and participants' contributions. In doing so, it creates a critical commentary that speaks out from the Asia-Pacific region, a locality framed by the aftermath of environmental disaster, onto a global stage.

Conclusion: Mobile Media, Intimacy, Place, and Critical Intervention

In this chapter, we have explored the role of mobile and screen media in understanding the relationship between art, politics, and the environment. As we have shown, mobile media, and camera phones in particular, are increasingly part of the practice of visual artists in the Asia-Pacific. This phenomenon holds although access to such devices and platforms is unevenly distributed across the region, and notwithstanding debates about their proper role in arts practice. This development, as we have stressed, needs to be understood in relation to the ways in which the very technologies and apps it engages are part of globally distributed corporate and institutional power relations that are supported by national agendas and consumer cultures. It is also, as for digital media more widely, part of the process through which climate change is produced, and is therefore riddled with the same ironies that beleaguer the development of ICTs in the region more widely.

Nevertheless, as we have shown, mobile media have a significant role in the making of critical and participatory arts practice. Art can play a vital role in providing alternative ways in which to think through critical issues. Through play and creativity, new options and actions can be invented. An understanding of this potential offers us ways to consider how environmentally responsible uses of digital media might be engaged in the mitigation of climate change through arts practice.

One of the chapter's examples explored a media project in post-3/11 Tokyo that investigated ways to empower participants. *Keitai mizu* deployed camera phone apps and Twitter to consider the ways in which intimate copresence and ambient play exist in and around mobile gaming within the everyday, and how the camera phone is a critical part of the environment as an embodied and affective experience of place. Such ways of bringing together localities of environmental disaster with mobile media arts projects, we argue, offer important models of how digital media can be engaged for critical intervention around environmental issues.

As we have seen in this chapter, smartphones provide ways for both everyday users and artists to engage socially, visually, and critically with multiple platforms. This sets part of the scene that we expand on in the next chapter, where by exploring the emergence of platforms as key sites for both media and art, we argue for attention to their relevance as both strategy and metaphor across media and art fields in the region.

5 Platforms for Public Engagement

In chapter 4 we discussed how online digital platforms create infrastructures and host representations through which people experience and engage with everyday and environmental issues, in particular with disasters. In this chapter, we shift the discussion to art and its relationship to the emergence of the platform as a concept and as what Massey (2005) has called an "event of place" through which artists engage publics across the region. We trace the rise of transregional spaces in contemporary art, such as biennials, community arts projects, and microspaces, in the Asia-Pacific, discussing three key ways in which they contribute to the making and articulation of critical perspectives on climate change within the region.

First we examine how platforms participate in shaping a new "cosmopolitan imaginary" of the Asia-Pacific (Meskimmon 2011; Papastergiadis 2012). Second, we focus on the critical attention being paid to the region's art practices and politics that has emerged with these events, and how these in turn are implicated in global mediated and economic flows. Third, we consider how these exhibitions, projects, and organizations can be understood as platforms that generate new ways for artists and publics to reimagine and experience art in the region, as well as the role of online public media in these exchanges. These platforms, we argue, not only represent regional identities, ties, and entanglements but also actively shape them.

The position of artists in this context is made complex, not least by the emphasis on temporal practices, collaboration, and social exchange, as well as the way in which it is marked by a cosmopolitan and global worldview (Meskimmon 2011; Papastergiadis 2012; T. Smith 2011a). This art world is an environment in motion, with the increasing circulation of artists and artworks often working transculturally. It resists any attempt to demarcate identity based on fixed ideas of boundaries that shape people and things. Focusing on art platforms not only offers us a key insight into this complexity but also offers an example of how such cosmopolitan places are emerging.

Platforms

As a discursive term used in art, architecture, politics, media studies, and marketing, and as a technical descriptor of digital systems, the platform has become a semantically flexible catchall metaphor in contemporary life. "Platform" is an increasingly popular term in contemporary art and events and has become a means through which to capture a multiplicity of practices. These include the naming of spaces such as Sàn Art (discussed in chap. 1); curatorial strategies such as Gwangju's Biennale Roundtable in 2012 (South Korea) and the Fifth Auckland Triennial in New Zealand in 2013; and defining networking practices and connections to sociopolitical realities, as developed through *Platform Seoul* (South Korea). The concept of platforms can be traced back to Harald Szeemann's Documenta 5 (1972) and Okwui Enwezor's Documenta 11 (2002), which used platforms as a metaphor to expand the "exhibition" across multiple events and sites.

The nomenclature of platform terminology has been popularized not only in the art world but also in the cultures of Web 2.0 and socially engaged online media practices. Gillespie (2010) has shown how the concept of "platform" has been used by online content providers such as YouTube to strategically position everything from technical services to the description of creative user actions and advertising content. He argues that these uses connote the concept—literal or metaphorical—of an expansive, level, nonhierarchical surface, hosting multiple activities or events (Gillespie 2010, 350). Platforms thus "anticipate" certain actions, implying a certain neutrality or egalitarian organization that provides an open or democratic space supporting users (individuals or corporations) (350). Such metaphors echo the rhetoric of Web 2.0 as a place facilitating user-created content, amateur creativity, production, and networking. Yet, as Gillespie demonstrates, the "comforting" ideas of free space and openness obfuscate the tensions and power relationships between the roles and actions of providers, users, and commercial media companies in the creation, distribution, and controlled delivery of online content. This slippery and strategic framing of platforms can also be applied to how contemporary art curation negotiates the relations of place in regional ecologies.

Contemporary biennials are a useful example through which to begin to understand these dynamics. As Terry Smith notes, there is a new recognition that contemporary art practice is *of* the world, which is particularly associated with the biennial. As he puts it:

As biennales have for decades attested, art now comes *from* the whole world, from a growing accumulation of art-producing localities that no longer depend on the approval of a metropolitan centre and are, to an unprecedented degree, connected to each other in a multiplicity of ways, not least regionally and globally. (T. Smith 2011b, 20)

Biennials and Regional Ecologies of Place

Biennials have a long history, dating at least as far back as the first Venice Biennale in 1895, and to even earlier models in nineteenth-century world expositions. However, they have more recently been distinguished by their increasing number, diverse curatorial modes, and the socially engaged and temporal forms of art production they display. These changes mark contemporary biennials as sites through which multiple elements of regional ecologies might be engaged with. While they have attracted much criticism for their imbrication in forces of neoliberal capitalism, such as their overblown budgets and supersize presentations and populism, as well as for their anthropocentricism (Williams 2014), they are more than just "mega-exhibitions" or elaborate spectacles. Biennials are places in which contested identities, the sociopolitical agencies of art, and its role in the political economy are played out—that is, they are "rewriting" regional identity. Even when biennials are sites for artists to oppose, they are still implicated in a form of interdependency. These events themselves can generate considerable attention through counteractivities. For example, an antibiennial exhibition was developed in Shanghai called *Fuck Off* (2000) (Merewether 2007), and protests grew up around the sponsorship of the Biennale of Sydney in 2014. Both examples attracted significant art world and media attention generating exposure for the both the anti-event and protest as much as the biennials.

The increasing rise in the number of biennials, triennials, and art fairs in the past twenty-five years across the globe has been well documented (Chiu and Genocchio 2010, 2011; Clark 2007, 2010; Filipovic, van Hal, and Ovstebo 2010; Gardner and Green 2014; Enwezor 2002, 2009; T. Smith 2011a). A significant portion of these have occurred in the Asia-Pacific, with more than twenty currently operating in the region, including Taipei 1992, Shanghai 1996, Gwangju 1995, Fukuoka 1999, Asia Pacific 1993, Busan 1998, Echigo-Tsumari 2000, Yokohama 2001, Auckland 2001, Guangzhou 2002, Beijing 2003, Singapore 2006, and Jogja 2009 (Biennial Foundation, n.d.). This has paralleled the rise of global art networks, the expanding international market for contemporary art, an expanded world art discourse, and a general increase in international art spaces, including online spaces. Biennials play a significant role in contemporary processes of globalization, acting as a counterforce to Euro-American hegemonies and other standardizing forces, and forging connections between local, regional, and global publics (Gardner and Green 2014).

The significance of biennials should not be overstated. A number of other transregional activities and events have also influenced development in the region, providing critical intraregional networks and identities (N. Taylor 2011). These activities also include publications such as *Modern Asian Art* (Clark 1998) and art journals such as *ArtAsiaPacific* (established in 1993). More recently, online organizations including, Art Radar Asia, ART iT, Media Art Asia Pacific (MAAP), and Arts Network Asia have provided critical conduits for accessing information on recent art practice, discussions, exhibitions, and publications in the region. While biennials have provided increased opportunities, they are in danger of being closed circuits owing to the limited number of artists and curators constantly being recirculated through them (Asia Art Archive 2011a; Clark 2007). Many of the organizational platforms emerging in the region are actively circumnavigating biennials and providing alternatives to this model of international engagement.

The so-called rise of Asian art in this context is part of broader social, economic, and political changes within the region. As authoritarian states transform through capitalist expansion schemes, art is becoming a critical tool in the political economy, and cultural hubs are being carved out in the region through major urban redevelopment projects. This can be seen as a variation of what Marc Augé (2013) refers to as a "proof of presence," whereby governments use art biennials and the architectural icons of cultural institutions to position themselves in a global political and cultural economy. Recent examples of development projects intertwining art, politics, and commodity in the region include Himalayas Center in Pudong, Shanghai (Tan 2008), West Kowloon Cultural District in Hong Kong (Hong Kong Economic and Trade Office 2013), and Gillman Barracks in Singapore (Seno 2009).

The increase in mega-exhibitions globally has also been supported by the expanding discourse on "world art" (Belting 2009; Belting et al. 2011; Elkins 2007; Harris 2011) emerging from discourses of postmodernism and postcolonialism, which dominated the art world in the 1980s and 1990s. While contemporary art continues to grapple with the tensions of an increasingly complex and interconnected globalism, questioning pseudo-transnational effects and acknowledging cultural multiplicities and identities, art institutions and organizations (including curators, artists, and scholars) are generating a different map. This involves moving away from East–West binaries or a (de)centralized Euro-American perspective and instead recognizing transregional identities. Recent high-profile examples include C-MAP, Contemporary and Modern Art Perspectives, at the New York Museum of Modern Art and the Guggenheim's UBS MAP Global Art Initiative.

Biennials and the Politics of Identity

It is easy to criticize biennials or reduce them to commodified spaces of art consumption, consequences of neoliberal forces, and spectacles driven by government-led urban development and the trends of the art market. However, such approaches limit our understanding of the role these events play in shaping the region's art. Biennials actively create new spaces and methods for artists to engage and amplify the contradictions and ambiguities of contemporary global ecologies. For example, the Asia Pacific Triennials (APTs) have been crucial in forming new connections between the local, regional, and global in a manner that emphasizes dialogue and interconnections, rather than universalism or complete fracture (Gardner and Green 2014). Curatorial selection and presentation in biennials also engage tensions and debates on artists' national and ethnogeographic identities and their framing. For example, the APTs' continual questioning of national identities and cross-regional interconnections shifts from earlier iterations that collated artists into national groupings (at least in the catalog presentations) to more recently focusing on local and transregional connections (such as the Mekong section at the 2009/2010 APT).

In the 1990s and early 2000s, collective approaches to national identity were critical in actively challenging and revising ideas of modernism and national identity by opposing Euro-American hegemony (Clark 2007). More recently, such homogenizing approaches have been retracted in favor of redrawing the map of contemporary art away from national frameworks toward transregional ones (N. Taylor 2012). As Richard Streitmatter-Tran (cocurator of the Mekong section at APT6 in 2009/2010) argues, the opportunities afforded through increased transregional exhibition opportunities have enabled artists—and here he is referring specifically to Vietnam, where state support for contemporary art is minimal—to operate outside national institutions (Asia Art Archive 2011a). The regional approaches of the APT have played an important role in establishing an identity that does not continually presume a centralized West against which all else is measured (N. Taylor 2011, 9).

Yet the APT can also reestablish centers within the region, such as strategically positioning Australia, and more specifically Brisbane (which was not previously recognized as an art center internally in Australia), as a key hub of the Asia-Pacific region through its "curatorial imagining" (Maravillas 2006). Likewise, it has also been criticized for "artificially" grouping artists together, such as the 2009/2010 APT presentation of artists from the Mekong region (Hajdu 2009).

While overt regionalism, rather than globalism, at biennials has been critiqued (Cruickshank 2006), it need not be at the expense of globalized perspectives. For example, the 2013 Singapore Biennale, *If the World Changed*, focused on regional practices in Southeast Asia, particularly those occurring outside the usual capital city centers, as a critical strategy. While the biennial is seen as an "international platform," the specific aim of Singapore 2013 was to "build a distinctive Asian identity" (Singapore Biennale 2013). Singapore's focus can also be understood in context of the government's aim of becoming a central regional arts hub.

Categorizing artists based on where they are from is also a problematic marking of identity, not only because of the increasingly nomadic life of many contemporary artists and artworks, but also because doing so foregrounds the ethnogeographic and biographic identity of the artist before any other reading of the artwork, including a discussion of aesthetics (Antoinette 2007). More recent biennial examples have tried to rebalance these readings, such as APT7 2012/2013, Singapore 2013, and Gwangju 2012 and Jogja 2013, by presenting artists according to the name (alphabetically) or subtheme, refocusing the emphasis on a broader "global" view and allowing for interconnections across the whole event. This action recognizes that identity is formed during mobility, and foregrounds the types of relations that shape the biennial as a place-event, as something always in process, in that "it is no longer where you are *from*, or even where you are *at*, which matters; it is more about the way we communicate with others" (Papastergiadis 2011, 73). In chapter 2, we identified how these connected relations are critical in forming ecologies of place. It is the *how* that is critical here, because it is actualized through screen cultures (screens are ever-present spectacles at biennials and through which many biennials are encountered online). Biennials are not just physical sites of exhibition; they are also symbolic places that contribute to the forming of regional identities.

In chapters 1 and 3, we outlined how the notion of the Asia-Pacific has been flexible in relation to the range of political, economic, and other discourses and interests that have strategically shaped its boundaries. For instance, in recent art discourse, "Asia" has been connected globally and extended in various ways. Biennale Jogja XII Equator #2 (2013), focused on collaborations between Asian artists and artists working in Egypt, United Arab Emirates, Saudi Arabia, Yemen, and Oman; and the 2010 and 2013 APTs included artists from the Indian subcontinent and western Asia. While these expansions are part of ongoing processes of transnational and intraregional connections and identity formations, they are not autonomous from market forces and trends that identify these areas as "emerging art scenes" (Hujatnikajennong 2013, n.p.).

Agung Hujatnikajennong, cocurator of Jogja XII, argues that this rhetoric betrays a Eurocentric perspective that aims to "appropriate patterns of production and distribution in the current global economy" (Hujatnikajennong 2013). The agency here lies with making these connections an explicit part of the biennial. Another tension related to representation and identity has been the tendency of discourses on the Asia-Pacific to emphasize Asia over the Pacific.

Like Asia, the Pacific is a contested category, encapsulating a vast array of cultures and geo-topographies, ranging from the west coast of North America to the southern Pacific region. As scholar Pamela Zeplin (2013) has noted, despite its vast size and diversity of language and culture, its popular image in the media in relation to rising sea levels and climate change, its significant migrant populations in Australia and New Zealand, and relationships created through colonial encounters, the Pacific remains a "curious omission" in many contemporary art collections of the Asia-Pacific or exhibitions of the global contemporary. There are a few notable exceptions, such as Queensland Art Gallery/Gallery of Modern Art (Brisbane), National Gallery of Victoria (Melbourne), and Auckland City Art Gallery (New Zealand). When it is included, the Pacific tends to be limited by an identification with fixed ethnogeographic categories, a temporality based on colonialist imaginings of modernity or curatorially packaged in popularized and politicized images of environmental disaster, tourism, poverty, corruption, and unstable governance (Zeplin 2013). Such staging continuously repositions the contemporary Pacific as the "other" in the Asia-Pacific. Furthermore, many of the contemporary works included in exhibitions or collections of the contemporary Pacific focus on artists residing in New Zealand or Papua New Guinea (Jolly 2007; Zeplin 2013). This does little to offer new perspectives on the Pacific diaspora and intraregional connections.

One of the factors that complicates examining the inclusion of Pacific work and its impact on forging an identity for the Pacific is the consideration of *who* is included and how their relationship to the broader region, including Asia, is represented. Artists and exhibitions, such as the 2012/2013 APT, which included a special cocurated exhibition of artists from Papua New Guinea, give visibility to the contemporary Pacific. They challenge demarcated binaries between traditional and contemporary practices in art and the categorization of the contemporary. However, the curatorial strategy of appointing a separate cocurator for this section and presenting it as an exhibition within an exhibition, with a separate listing of artists on the APT website, could also be understood as differentiating the Pacific as something apart, the "other" within the biennial event.

Attempts to address the imbalances of Pacific representation include the 2006 counterexhibition The Other APT in Noumea and the online component of the 2008 Biennale of Sydney, which sought to identify connections between artists from Pacific, Indigenous (Australia), and Asian backgrounds. Community programs such as the Contemporary Pacific Arts Festival (Melbourne, 2013) and Edge of Elsewhere (part of the Sydney Arts Festival, 2010–2012, organized by Campbelltown Arts Centre and 4A Centre for Contemporary Asian Art) explicitly foreground Australia's relationship to Asia and the Pacific to reconsider community engaged practices as transregional and interconnected. Edge of Elsewhere demonstrates how "local" place, in this case the western suburbs of Sydney, is formed through regional migration (Andrew et al. 2010). This does not divest the local of its importance or particularity, or likewise its national identity, but rather allows for a reimagining of the connections between identity, place, and community as transregional, connected through the material, virtual, and imagined.

Platforms as Places of Collision

The Ninth Gwangju Biennale, *Roundtable*, is a recent example of a new structural approach for understanding and activating the diversity of the Asia-Pacific. Aiming to critique linear narratives and a universalizing approach, the curatorial strategy of this event was based on the appointment of six cocurators, all women who were working actively in the region: Nancy Adajania, Wassan Al-Khudhairi, Mami Kataoka, Sunjung Kim, Carol Yinghua Lu, and Alia Swastika (Gwangju 2012). While the appointment of multiple curators had the potential for a lack of cohesion, and arguably evidence of that lack can be found in the wide range of artworks and artists selected and in the breakdown of the exhibition into six subthemes, it also constructs a model of curatorial practice based on relations of collaborative participation and exchange.

The metaphor of the roundtable references both a type of political gathering and a traditional Korean *duriban*, a communal eating structure (Gwangju 2012). The emphasis was on a "conversational" approach, which allowed for flexibility and aimed to encourage collaboration among the curators. The exact nature of their collaboration in practice is not clear, and it remains uncertain if it emerged as an entangling of influence or decisions on the ground, or whether it remained aspirational, with the curators maintaining a level of autonomy. The tensions around contemporary art collaboration are analyzed in more detail in chapter 7. The achievement of *Roundtable* was, however, to critique collaborative approaches in which differences are erased, and instead to foreground the frictions and fissures arising during connection as a form of activated fusion in which "the act of curation is thus a synthesis—a series of collisions that perhaps leads to transformation" (Gwangju 2012).

"Collision" is a useful term, suggesting both an aleatoric process and one that is ultimately transformative rather than a destructive dead end. It echoes Arjun Appadurai's (1996) writing on the disjunctures that occur through the social life of things, including art, as they circulate and collide with imagined identities in globalization. Understanding collaboration as collision creates a framework through which art can be reimagined as a fluid set of relations connected to the environment. This framework amplifies the sense that art is emplaced *in* the world. It enables us to reconsider the role of art not just as representing place, including regional identity, but rather as being embedded in the process of the region's ecology. As noted by Terry Smith, contemporary art is *of* and *from* the world. It creates encounters, exchanges—spaces in which cultural difference, local identities, and regional connectivities are being formed.

The roundtable approach at the Gwangju Biennale used platforms to promote the curatorial management and exhibition approach as democratic, challenging the central or singular focus of many similar events that have only one curator or artistic director. *Roundtable* was specifically used reflexively to refer to the political roundtable, to "engage with the practice of the collective as a non-hierarchical organizational alternative" (Gwangju 2012). This approach was designed to open up to, rather than limit the recognition of, the different types of art practices occurring in the region by using the local knowledge of the curators. In this sense, the biennial was designed very differently from the fly in/out model whereby an international curator would be commissioned. The rationale for this approach was that such plurality in curation is *more* flexible and is distinct from the singular artistic director approach.

More cautiously, one might argue that having multiple curators in fact increases a multitude of "singular" visions that at worst offer only chaos and loss of direction. However, it is difficult to generalize, since such locally organized events will always be subject to specific circumstances. Another example of how a nonhierarchical organizational structure was developed is *The Lab* at the 2013 Auckland Triennial. This was an open-ended curation-within-curation model that aimed to emphasize the exhibition as a creative and social site of production, open to interaction and change, and to access local networks and knowledge:

The triennial is not only an exhibition but a machine of knowledge, a machine of experiment. ... Another important aspect is how to engage the triennial ... and also contemporary art in general—with local communities to create another social relationship and a social network. (Hanru and King 2013, 20)

The emergence of these ways of going beyond and critiquing the singular curatorial model raises the question of the extent to which such collective approaches can enable synthesis or connection across platforms. One way to avoid merely "rebranding" conventional curatorial methods and exhibition structures while continuing them in practice has been to shift the physical places in which art is encountered. For instance, Sunjung Kim (2014) used the concept of platforms for the festival *Platform Seoul* 2006–2011 as a means to connect art's production with place, and to provoke questions about art's sociopolitical agency. She wanted to move beyond the biennial model to create something with long-term sustainability and criticality.

For Kim, a platform was a social space rather than an actual physical site. Rather than solely focusing on exhibitions, *Platform Seoul* comprised elements including exhibitions, artist talks, public lectures, commissioned works, and seminars, which were designed to respond to different sites in Seoul (a different site and theme were selected for each festival). These elements were designed to be interactive, "to conjure the synergetic platform as a social and artistic field of transferability, conversion and openness" (Kim 2014), and one that worked with site. *Platform Seoul* can as such be seen as a place-event—mobile, evolving, connected to the environment, marking out identity. It is material, social, and symbolic, and these elements constitute a type of relationality.

Similarly, Gwangju foregrounds the idea that biennials go beyond simple representation in that they are shaping place, not just representing it. This sentiment was echoed by Hou Hanru, curator of the Fifth Auckland Triennial, *If You Were to Live Here …* (2013). Hanru critiques the biennial as a representational tool, arguing, "It's about creating a site of production of various ways of inhabiting, living, producing knowledge, imagination and conversation" (Hanru and King 2013, 19). Rather than biennials being some kind of "spectacle" that is separate from the social, political, material, and cultural relations of place, they are part of place, a form of "coexistence" with contemporary life that marks them as distinct from previous forms (T. Smith 2011a, 13).

This raises two critical points. The first concerns defining the role of biennials in constructing place-events by understanding them as platforms. This occurs through the interaction of identity and the production of social relations, but also through the ways in which biennials form temporal communities and temporal physical spaces for the duration of the event. As events, they form specific "publics" who participate in and generate the event. How these publics interact with the work, and whether the art sacrifices a sharp sociopolitical critique for popular spectacle, is a critical question.

Gardner and Green (2014) identify a new form of "intimate public" emerging through these events, albeit one that emerges, in their view, from a new "popular playground" form of spectatorship wherein the political aesthetic of relational practices, the rhetoric of "labs" and "utopias," and the vast scale and budgets of biennials drift away from critical democratic politics into populism. However, they highlight what art events like the APT critically contribute: "the presentation and production of publics that are able to bring together the very local (through hands-on experiences of new technologies), the regional (the Asia-Pacific), and the global (renowned artists, city boosterism, globally networked technologies) in ways that aspire to be neither homogenizing nor didactic nor foreclosed" (Gardner and Green 2014, 33).

Second, biennials foreground the role of the imagination and art practice in mediating the connection between local and global relations, between representing and shaping place. One aspect of this includes the role of biennials in the quick waste consumption culture of neoliberal capitalism. Biennials are large events that require not only significant infrastructure and material products but also, given their temporal nature, large amounts of capital that can be invested in one-off, site-specific works that are torn down at the end of the event, rarely to be seen again. Art, as we have argued, emerges from the same processes and material, social, and symbolic relations that constitute the environment. It is not something autonomous and separate, a representation of material consumerism, but rather part of it.

Biennials also have a role as sites of social knowledge construction. This is enacted through commissioned work, site-specific events, and artworks that highlight the relationality of place and social aesthetics. Nikos Papastergiadis identifies this role in the 2002 Gwangju Biennale curated by Hou Hanru, which also focused on the biennial as a "platform for social relations" (Papastergiadis 2012, 156). Here art was seen as a collaborator in the production of place, working in conjunction with the site, event, audience-participants, and, as Papastergiadis highlights, the imagination. This is "the means by which artists participate in the mediation of a cosmopolitan imaginary" (159).

Cosmopolitanism is an increasingly popular framework through which to understand this relationality and emplacement (Meskimmon 2011, 2013; Papastergiadis 2012). "Cosmopolitan" is a thorny term, particularly in the Asia-Pacific, owing to the term's roots in nineteenth-century European colonialist ideas of a universalizing and privileged "world citizen." However, contemporary scholars such as Papastergiadis are keen to avoid these connotations and to instead use the notion of cosmopolitanism in relation to actions that express a sense of global community and as a way to negotiate cultural differences (ethnic, class, gender) and to acknowledge and account for the "disjunctive diversity" in a worldview (T. Smith 2011a). Papastergiadis is interested in the politics of a "reflexive hospitality" in art, including the self-awareness generated through exchanges and collaborative production. He suggests how hospitality can easily slip into positions of easy conviviality or conversely put the focus of art on the ethical practices of collaboration and exchange. However, it is through collaborative practices that Papastergiadis argues that artists "become more intimately involved in the production and mediation of a cosmopolitan imaginary" (2012, 159).

In reconsidering art's political agency in the context of socially engaged art practices, Papastergiadis argues that art is a social and political aesthetic, not a representation of it, and this is where art's agency is activated (2012, 14). These ideas are echoed by Nancy Adajania, one of the six Gwangju curators, who emphasizes that transcultural experiences are performed through art, and that the political potential of art resides not in "picturing potential" but in creating conditions (Venkatraman 2012, n.p.). There is also, however, a need to acknowledge the ways that materiality and technological dimensions of art affect the world, and to do so takes us back to the questions about waste and environmental crisis that underpin this book. Biennials are as much productions of capitalism, contributing to a consumer-driven curatorial approach and throwaway material culture in art, as they are sites in which new relations of place are articulated.

Furthermore, despite the criticisms of the populism of biennials, they are still arguably engaging limited audiences, composed of specialized art publics but also of privileged middle classes who can travel to and attend such events. Likewise, the artists and curators who participate are also those who have access to travel and to niche global art networks so as to be invited to participate. Again, inclusion in the forms of cosmopolitanism that biennials generate is therefore not entirely open entry but privileged and specialized. While the recuperation of political agency in aesthetics is difficult to fully resolve, what events like biennials do create, however, are opportunities for these tensions to unfold and play out. They also offer a series of examples of the potential of regionally constituted cosmopolitan art platforms in exploring how public art in the Asia-Pacific can participate in democratizing processes that might contest and address questions about climate change.

Marsha Meskimmon identifies cosmopolitanism as a "precarious ecology," "a process or state of relations between subjects, objects and their environment" that is actively being composed and produced always with the risk of being undone (2013, 40). Although Meskimmon acknowledges the "agency of art as a world-making practice, rather than a representation of the world" (40), unlike Papastergiadis her focus is equally limited by her concern with the materiality and embodied affectivity of *art objects*, rather than with more socially engaged art practices. However, her acknowledgment of cosmopolitanism's development as ongoing, mutable, dynamic, and, importantly, precarious is significant in shifting the focus away from representations. Meskimmon argues:

Cosmopolitanism is neither predetermined in content, nor in form; instead, it is understood as a carefully poised system of relationships that are open to difference, managed through intersubjective generosity and negotiation with others, and constantly changing as the material constraints of the environment evolve. (2013, 40)

By moving the focus toward how artworks (but this can equally be applied to art events, projects, or organizations) create the possibility for "the interpellation of cosmopolitan subject-positions and interrelationships" (Meskimmon 2013, 41), we can see how the new types of art spaces emerging in Asia actively foreground art's agency in shaping cosmopolitan encounters, social production, and knowledge construction. They operate as material, social, and symbolic spaces—both online and offline—that forge regional art ecologies and engage in transregional networking. As such they also begin to form platforms through which art may help to foster regional responses to climate change and environmental issues.

Platforms for Engagement: Emerging Spaces in the Region

Alongside the growth of biennials, significant networking opportunities among artists, curators, and arts organizations have emerged in the region, providing new forms of material and symbolic sites for contemporary art. These new sites can be understood as "post-alternative" spaces and should be distinguished from the local avant-garde activities and artist-run initiatives of the 1960s and 1970s. While they offer alternatives to mainstream institutions and major biennial events, they do not necessary hold "oppositional" positions, as they may have the support of larger, established institutions, government funding, private partnerships, and sponsorship. They often provide a critical role mediating relationships between local artists and external organizations and events such as biennials. They are shaped by the specific social, political, and economic forces of the region and locality; contribute significantly to the rising contemporary art discourses in the region; and represent critical nodes in a regional meshwork. Examples include Utopia@Asialink (Melbourne), Asia Art Archive (Hong Kong), Soundpocket (Hong Kong), 3331 Arts Chiyoda (Tokyo), Sa Sa Bassac (Phnom Penh), Ruangrupa (Jakarta), 31st Century Museum of Contemporary Spirit, Station

(Chiang Mai), Samuso: Space for Contemporary Art (Seoul), and Vitamin Creative Space (Guangzhou).

These new spaces for contemporary art are not just exhibition sites; they offer a range of public programs and work through many different forms of funding, including foreign grant support from organizations like the Asia Foundation, the Japan Foundation, and the Goethe Institute, as well as private funders and patrons, and partnerships with overseas galleries and institutions. While this type of funding can be both precarious and volatile, subject to external controls, it also means that these organizations have the potential to be responsive and adaptable, unlike larger, more established institutions that plan their programs out years in advance. An example of one of these new spaces is 3331 Arts Chiyoda's WAWA Project. This project is focused on regeneration and recovery for the Tohoku region and was able to set up in the wake of the 2011 earthquake and tsunami disaster (see chap. 8).

Other artists in Japan are using online and print publication spaces as a method for community engagement and to provide sociopolitical critique in a post-Superflat and post-Fukushima Japan. One example is the art collective Chim↑Pom, which occupies digital online platforms as a means of art production and distribution (http://chimpom.jp), employing video, performance, and installation to engage in sociopolitical critiques. More recently Chim↑Pom's artists have extended their activities to also focus on curation and publication. They became well-known through their bombastic, Dada-esque urban interventions, such as their provocative installation of images of the Daiichi nuclear power station at Fukushima. This was pasted over Taro Okamoto's mural *Myth of Tomorrow* (1968–1969), which addressed the bombing of Hiroshima, at Shibuya Station. The act quickly attracted attention on social media sites such as Twitter (PBS Frontline 2011), generating speculation about the installation's origins. The work was removed within twenty-four hours. Despite its short life span in the station, the action carved out a significant place, particularly online, in discourses addressing the events of 3/11 and Japan's use of nuclear power. The group later posted a video on YouTube of the act of installing the work and titled the piece *Level 7 feat. "Myth of Tomorrow"* (Chim↑Pom 2011a), which became a work in itself documenting the act.

Other projects, such as *Real Times* (Chim↑Pom 2011b), comprised performance and video art pieces based on the group's entry to the Fukushima Daiichi Nuclear Power Plant a few weeks after the events of 3/11. A series of deliberately staged and documented performances engaged with the actual physical environment of the plant at a time when the media were supposedly filming from kilometers away and touching up the images to make them look as though they were filmed from more intimate

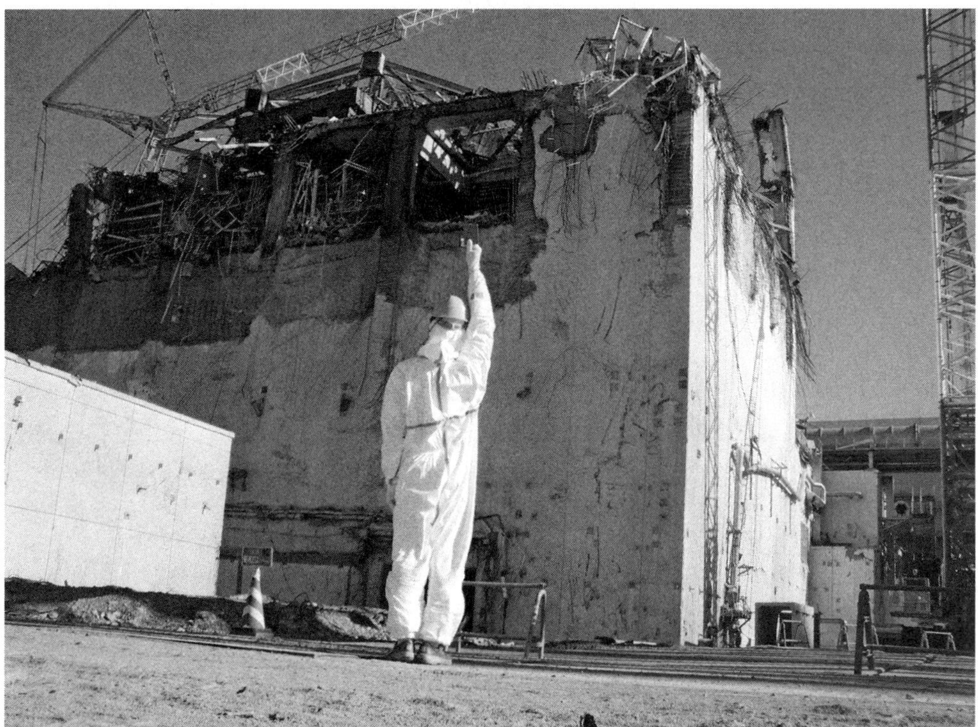

Figure 5.1
Chim↑Pom, *Red Card* (2011). © Chim↑Pom. Courtesy of Mujin-to Production, Tokyo.

positions (PBS Frontline 2011). The work engages with ideas around public discourses and the role of state and commercial news media in providing access to the ongoing effects of the earthquake and tsunami.

The work emerged at a time when people were not only wary of government rhetoric about "safe" levels of radiation but also reconsidering the social, political, and environmental role of art. Chim↑Pom not only wanted to experiment with creating works, such as the ikebana arrangements of flowers and flora gleaned from the restricted zone around the plant, that represented the disaster's environmental effects, but also to participate with local communities in rebuilding their material infrastructure.

Chim↑Pom is part of a shift toward socially engaged art practices and new types of places being occupied by art. This shift is partly driven by economic changes as artists continually seek out new sites in which to exhibit and practice. An example of this is *K-I-S-S-I-N-G*, a work by Chim↑Pom that was exhibited at The Container gallery, Tokyo. The Container is an actual shipping container located inside a (functioning) hairdressing salon. It is a type of alternative microspace common around the region, partly due to high rents in areas such as Tokyo and Hong Kong, but also as a way to bring art into new urban spaces and to use existing infrastructure rather than re-create a new gallery space.

Another example of a multifaceted organization dealing with alternatives to typical building and exhibition infrastructure and gallery space is Utopia@Asialink. Utopia originally modeled itself on Manifesta and intended to be a biennial program for visual arts; however, it restructured because of questions about the need for yet another biennial event in the region and the significant associated costs and infrastructure. As an alternative, it was set up in partnership with institutions and key art figures in Tokyo, Singapore, Seoul, and New Delhi to create a transregional platform that could use existing infrastructure (galleries but also events) and be mobile and flexible—an "incubator" for contemporary art focused on collaborative engagement and capacity building (http://asialink.unimelb.edu.au/arts/utopia).

Asia Art Archive (AAA), Hong Kong, supports local artists and professionals through a significant physical library collection that covers art and exhibitions in the region, workshops, and lectures, but AAA also actively creates connections regionally and internationally through digital archives, visiting artist residency programs, and art commissions (http://www.aaa.org.hk). These objectives and activities, local and regional, work together to provide a broader context for establishing a critical contemporary art discourse. AAA is a central hub, both physically and via public media such as its online database that provides access to a vast collection of art-related material, and it plays a significant role in mediating relationships between locally based artists and regional or international opportunities.

In contrast to these examples, in Vietnam there is minimal infrastructure and institutional support for contemporary art. Independent artists are marginalized relative to state-approved academic and traditional arts practices, and many contemporary art galleries rely on foreign funding to survive. One example is Sàn Art in Ho Chi Minh City. Established in 2007, Sàn Art is an independent, artist-run organization (http://san-art.org). As well as a physical gallery and library, Sàn Art is a mixed-use space shared with the Propeller Group (http://www.the-propeller-group.com), an artist collective that creates multimedia projects using the modes of contemporary screen culture, primarily television advertising, and the Internet to engage in sociopolitical critiques and mediated representations of Vietnam, such as the work *Television Commercial for Communism* in 2011.

Sàn means "platform," and although creating a physical gallery space for contemporary art is part of Sàn Art's aim, its reading room (which houses a significant collection of contemporary art texts) and public education activities are vital to fostering opportunities for artists and the public to engage with contemporary art discourse. Sàn Art also hosts artists' talks, public workshops, and a residency program. The organization aims to create a space for contemporary art discourse in an environment in which state support is limited and heavily regulated. Contemporary Vietnamese artists are seeking to break from the standardized collective art forms that are products of Vietnam's modern national identity. Sàn Art creates a place in which artists can produce these new identities and engage with an international contemporary art context.

Generating and participating in art discourse and education through discussions, providing access to readings through physical and online libraries, and organizing workshops are critical roles that many organizations and artists have played in supporting contemporary art practice in Vietnam. Other examples in Ho Chi Minh City include Richard Streitmatter-Tran's Dia Projects (http://www.diacritic.org), and Zero Station (http://www.zerostationvn.org), an artist-run space set up by Nguyen Nhu Huy. Such approaches explicitly recognize and promote art as a form of knowledge production and acknowledge the significant sociopolitical role of contemporary art in shaping public discourse and defining the region.

Galleries and spaces such as Sàn Art and the Asia Art Archive also operate as key meeting points for local artists, curators, academics, and visiting internationals, playing a vital role in creating and maintaining regional and international networks. They are hubs, part of a regional *meshwork* (Ingold 2008). They actively participate in shaping critical arts discourse and practices locally and in forging connections internationally and regionally. They can be understood as part of a meshwork because they are not just about mutual relations between two points—that is, they are not fixed or discrete nodes in a delineated network of things—but rather involve more interpenetrative relations, defined through movement (Pink 2011).

These art spaces can be understood as evolving ("precarious") ecologies that unfold across different platforms, both physical and online. The online–offline relationship distinguishes them from other major art institutions in the region, especially in their focus on promoting contemporary art practice rather than more academic or state-sanctioned art, as well public engagement and their role as education spaces for contemporary art discourse. They focus on strategic self-awareness in art as a form of knowledge production (including of its political and social role) as much as on aesthetic production. For institutions in Southeast Asia in particular, this is something that has been lacking or absent due to authoritarian and conservative approaches by the state to the arts.

Public screens—whether mobile media or digital online spaces—also play a significant role in the accessibility, distribution, and functionality of these art spaces, as examples such as the work of Chim↑Pom demonstrate. In particular, local public engagement and international awareness of these spaces are accessed through online media. This does not mean that physical places, such as gallery exhibitions or workshops, are no longer important; indeed, they are vital in supporting the practice of art, stimulating local communities of knowledge, and providing a space to *go to* and encounter the material forms of art. However, the online is increasingly playing a significant role in their public profile. While physical art places remain local in material form (whether gallery space, office space, or face-to-face artist talks and workshops), their outreach and how they are encountered are much broader and increasingly facilitated through screen spaces, such as Asia Art Archive's provision of online documents and resources or Planting Rice (Philippines) (http://www.plantingrice.com), an online project aimed at disseminating information and discourse on Filipino contemporary art.

In addition, online spaces facilitate a range of participatory practices. These range from Soundpocket's online library of "donated" sound recordings, a community project engaging with local sonic topographies (http://www.thelibrarybysoundpocket.org.hk), to the use of online media as an artistic platform for distribution and critique, such as the Propeller Group and Young-Hae Chang Heavy Industries (YHCHI) (http://www.yhchang.com). The online is not a secondary or "additional" space for these organizations; rather, it is one of many sites of engagement.

The pressure on art to represent—to be a platform place from which positions are articulated—is parodied in Young-Hae Change Heavy Industries' work *It's a Pleasure to Be Here: We Have Nothing to Say*, which they produced as part of their residency at the Asia Art Archive in 2011 (Asia Art Archive 2011b). YHCHI uses flash technology to create text- and sound-based works that mimic the vernacular and acontextualized language and aesthetic of the Internet. This particular work humorously parodies the expectation that contemporary artists, whether creating site-specific works or participating in residency programs, will somehow offer privileged insight into those specific places even when they may just be engaging in a fly in/out model of residency (such pseudo-ethnographic models are critiqued by Hal Foster; see Hjorth and Sharp 2014), and then to repeat the same process at the next event. YHCHI parodies the convivial and politely positive tones that seem to often frame the social relations of contemporary global artwork and the smoothing over of political difference using digital voice technologies, such as the seemingly "anonymous and impersonal" tones of user interfaces.

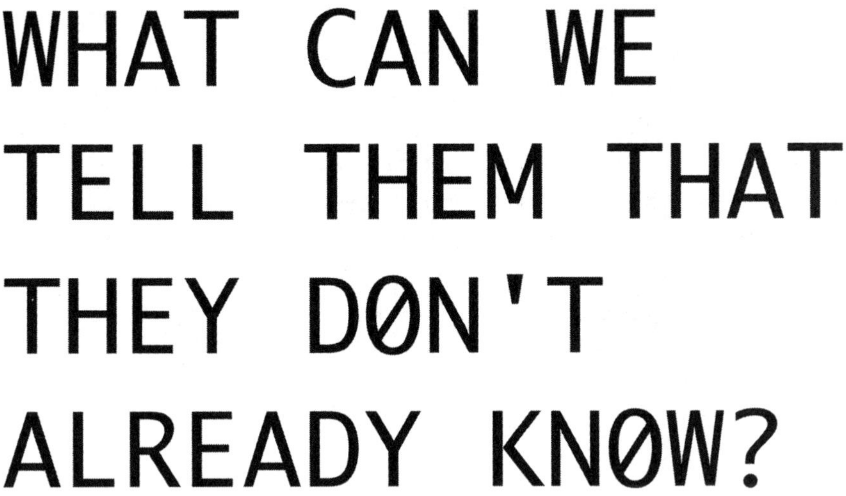

WHAT CAN WE TELL THEM THAT THEY DØN'T ALREADY KNØW?

Figure 5.2
Young-Hae Chang Heavy Industries, *It's a Pleasure to Be Here: We Have Nothing to Say*, 2011. Courtesy of Young-Hae Chang Heavy Industries.

The multiple activities of organizations such as Sàn Art, 3331 Arts Chiyoda, and Asia Art Archive can be understood as "platforms." Gillespie's analysis of how platforms "matter as much for what they hide as for what they reveal" (2010, 359) is a useful way to approach the new types of art spaces emerging in the region. These spaces position themselves in contrast to conventional gallery, museum, and exhibition spaces, and they operate multiple spaces (online and offline) to deliberately focus on art practice as social and knowledge production. The use of online portals, databases, and emphasis on artist talks, symposiums, and workshops, mark organizations such as Sàn Art, 3331 Arts Chiyoda, and Asia Art Archive as distinct spaces for contemporary art.

Conclusion

Contemporary art practice in the region is actively shaping an ecology of place and identity, forging connections and amplifying tensions. Biennials, transregional projects, and the new arts organizations emerging in the Asia-Pacific do not operate in isolation from one another but rather are part of an evolving meshwork. To understand this dynamic, it is therefore important not just to examine biennials and urban redevelopment, which have tended to dominate the discussion of changes to the art world in the region, but also to consider organizations and projects as platforms and place-events that define the ecologies of the Asia-Pacific.

The popularity of the term "platform" (alongside others such as "incubator" and "lab") points to approaches that foreground the types of mutable and transformative connections (and conflictive disconnections) that shape our experience of place. While the term can easily slip into rhetoric, it provides a vantage point from which to untangle the difficulties surrounding the shaping of a "cosmopolitan imaginary." In particular, how artists and arts organizations are using online screen media occupies a critical role in shaping different types of connections beyond the exhibition or festival site.

6 Art, Critique, and Climate Change

In the previous chapters, we outlined how, in a context where climate change is increasingly acknowledged as a scientific and experienced reality, new relationships between art, screen media, and environmental issues are emerging. Now we focus on a specific set of regional contemporary art practices emerging in the Asia-Pacific that are associated with an ecocritical approach.

Ecocritical theory within academic scholarship and practice-based and culturally specific forms of ecocritique have developed as challenges to normative models of nature and the serious environmental problems that accompany them. Such work has emerged both from Western scholarship (e.g., Braddock and Irmscher 2009) and, more recently, from Asian perspectives (Estok and Kim 2013). Ecocritical debates have shown how commodity culture often privileges pristine images of nature, refiguring nature as an object of desire rather than as a constituent of the world we live in. In this sense, images of "nature," such as those of water discussed in this chapter, have become a complex genre in a highly contested field of representation. Ecocritical perspectives show how everyday lives are actually enmeshed in material ecologies of which nature is a part, as opposed to conceptualizing nature as something separate that can be consumed or controlled.

Indeed, ecocritical perspectives have posed a persuasive critical challenge to the dominant claims of such representations, both through writing and through the materiality of art practice. In this chapter, we build on this perspective to examine how art is adding to the ways in which nature and the region can be imagined. In doing so, we consider three contexts: the relationship between art and urban ecologies in the region; how the region's cities are imagined locally, regionally, and globally; and how new imagined communities and publics have been constituted through critical arts projects that explore the theme of water and its environmental consequences across the Asia-Pacific.

Art-ificial Nature: Art and Urban Ecologies

The work of the artists discussed in this chapter can, and should, we argue, be considered part of emergent narratives of global, regional, and local imaginaries about climate change. However, we qualify this point by saying that it is unusual for regional environmental art addressing issues of urban pollution to extend beyond local sites. Artists in the Asia-Pacific are creating versions of this imaginary that are localized: their work can be seen as responding to the suffocating levels of air pollution in Asian megacities, or to the critical, unintended consequences of important regional energy projects such as the large dams and nuclear power plants we discuss later in this chapter. However, the local is never purely local. As our examples here show, cultural and civic engagement in local ecologies is increasingly becoming a way in which publics are engaging with the aesthetic qualities of the region's cosmopolitan cities. Projects such as Seoul's City Gallery Project exemplify this trend; the controversy, critique, and new visualities surrounding Cheonggyecheon Creek in Seoul demonstrate how such processes can play out.

The uncovering of Seoul's Cheonggyecheon Creek, which was covered for decades by roads, can be interpreted as a project through which the urban was brought together with nature. The controversial $900 million project originally attracted much criticism but subsequently became a key tourist site and urban imaginary. One look at the millions of daily camera phone images of Seoul demonstrates how prolifically Cheonggyecheon features (Lee 2009; Wong and Yuen 2011).

Similar ways of marrying urban design with nature can be found in locations such as Singapore (Hudson 2014). As Chris Hudson has discussed through public and urban art and architecture projects in Singapore, "greening" often necessarily involves the representation of nature within a focus on material sustainability. The growth of vibrant creative cities in the region has been accompanied by the role of nature and greening design. As Gang Chen (2010) notes, the notion of the eco-city, which provides an innovative yet pragmatic approach to designing, building, and operating cities, has become one of the key models in the region. This agenda has seen the creation of many public artworks that comment on the role of nature, whether aesthetically, literally, or ideologically. Most of these large-scale works are in parks, and some, such as *Living Light* (2009), for example, have an interactive capacity.

Living Light is based on an illuminated geolocative platform within a domed, treelike structure that has twenty-seven panels linked to twenty-seven air-monitoring stations across Seoul. Every fifteen minutes, these panels light up in response to the relative levels of air quality throughout the city. People can also read the quality of the air in

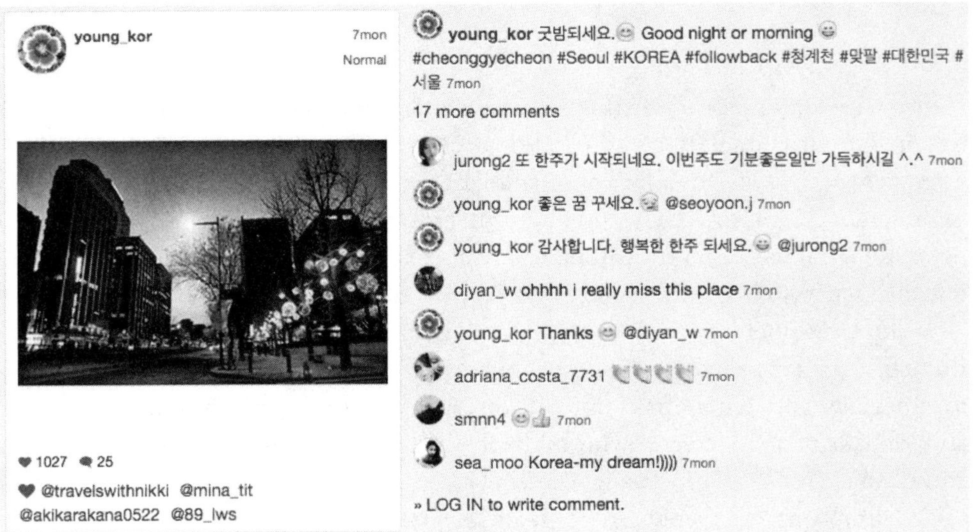

Figure 6.1

One of the thousands of Instagram images hashtagged #Cheonggyecheon.

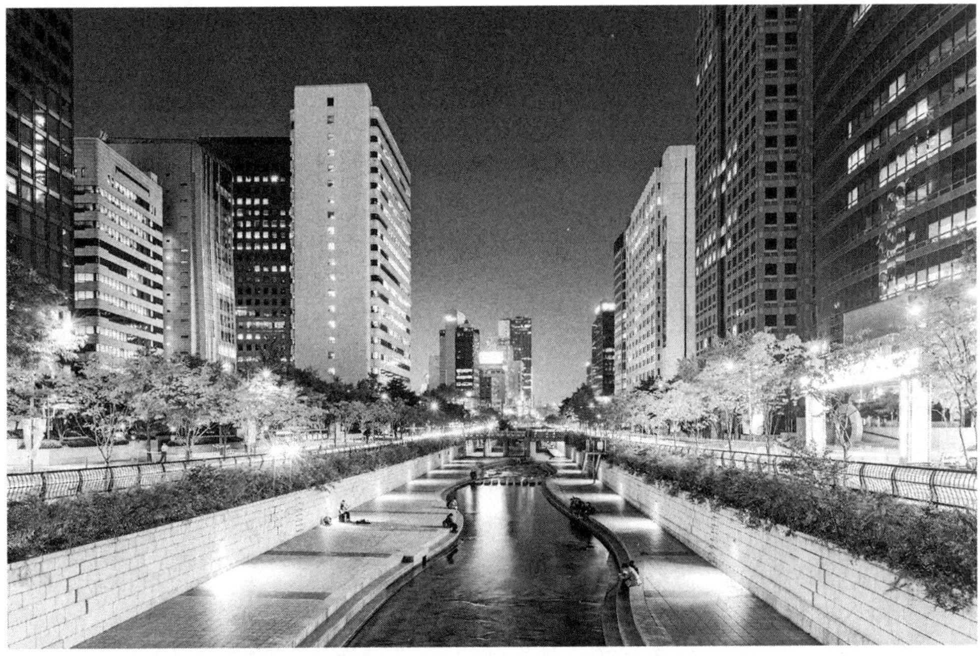

Figure 6.2

Cheonggyecheon Creek, a remade creek in the center of Seoul, plays a key part in reimaging the city's urbanity. (Photographer unknown.)

their own urban sector by texting their neighborhood code into the relevant panel, which lights up and sends the information back to their mobile screens. This is another, more benign, instance of how information about local ecosystems, such as the vital information communicated in crises wrought by extreme weather events, has extended the ontological indices of the intimate publics of mobile screens in ways that incorporate material environmental change.

Living Light offers an interesting example of how what people feel as they move through their environments can be connected to data as art. The work thus brings together the representational format of the art installation, and the data that it uses and sends to mobile screens, with the embodied sensory experience of being in and part of an everyday environment. These relationships between climate as science, and the environment as experienced, form an integral aspect of the ways that contemporary art, brought together with the affordances and qualities of screen media, can affect how people understand and engage with everyday environments and climate change.

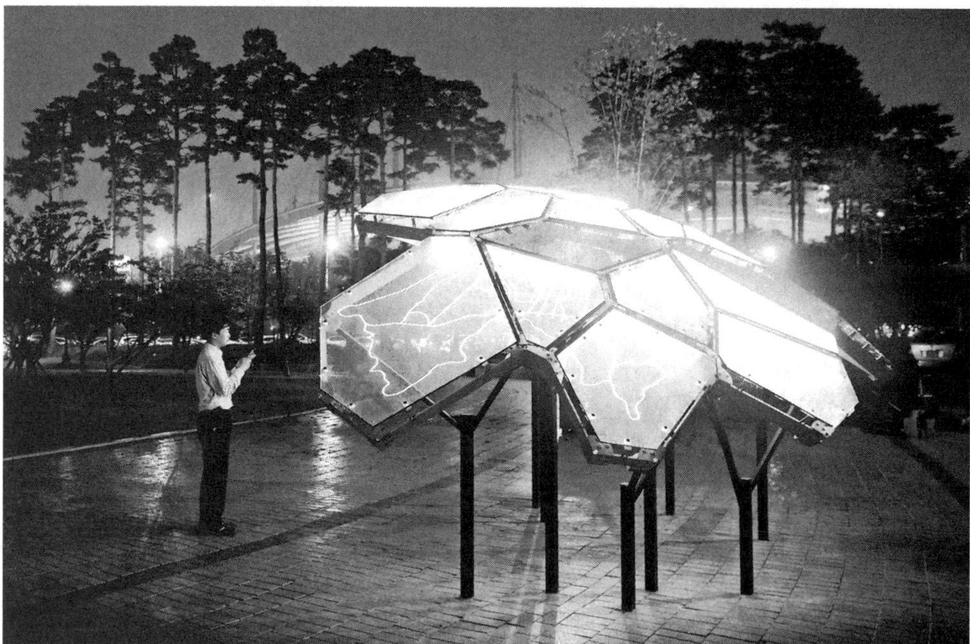

Figure 6.3
David Benjamin, *Living Light*, Seoul, 2009. Courtesy of the Living.

In Seoul there has been a concerted effort to reduce air pollution in the city by introducing gas-fueled buses and low-emission cars. By tapping into mobile and geolocative media, *Living Light* was designed to encourage public agency in this civic plan. The Chinese government is also acutely aware of the need to reduce air pollution in major cities, though not quickly enough for a dissident artist like Ai Weiwei, who in 2013 appeared on Twitter in a gas mask to protest the dangerously high level of air pollution in Beijing. There are also popular cultural attempts to address the problem; for example, Sonalika Jain, a schoolgirl from Shanghai, won the Eco-Art China Award in 2013 for her recycling of waste into art. This award was promoted by the Chinese Touchmedia Eco-Art project.

Developed in 2003 by Micky Fung, the Touchmedia company adapted taxis to carry interactive touch screen technology for advertising to middle-class urban consumers. This proved to be a lucrative platform for advertising while also bringing the company numerous government and public service awards for its philanthropic public projects, including the "Eco-Art China" campaign that the company initiated in 2009. In 2011, Touchmedia joined with the official Shanghai Environmental Protection Bureau to launch its "Play with Garbage" competition. This competition reached a massive public of over a hundred million people. This extensive taxi public was given comprehensive information about recycling as a means to combat pollution. People could play interactive screen quizzes while traveling in the cabs and were also able to vote on the touch screens for their favorite artworks constructed out of garbage.

Though currently based in several Chinese cities, including Beijing and Hong Kong, Touchmedia's Eco-Art campaign also extended national cultural borders by displaying artworks from twenty-eight countries, with recent winners chosen from New Zealand and India, as well as China. The Eco-Art China website continues to recruit participants from across the region. The voluntary organization Greening the Beige (GtB) is another such example of international dialogue, as its founders, though not originally from China, now live in Beijing, where they focus on public awareness campaigns through art, addressing the problem of air pollution in the city.

Although Australian cities are also affected by air pollution, its effects are less extreme than in the much larger Asian cities. Cities in Australia, however, are becoming more vulnerable to increasingly intense heat waves and to the "urban heat island effect" in which extreme conditions are exacerbated by a buildup of heat due to urban density. This context leads to a series of issues that affect people as they live their everyday lives in urban Australia. Issues include the generation of peak energy demand for air-conditioning, which has implications for energy supply infrastructures and economic costs, as well as creating risks of power cuts that affect vulnerable citizens during heat waves (Strengers et al. 2014). As we have noted, even in the cautious language of the most recent IPCC report, more extreme summer temperatures are likely to occur in Australia due to the consequences of global climate change (IPCC 2013).

This critical perspective on climate change became the focus of an artwork in Sydney when the Australian artist Allan Giddy adapted two big construction cranes in a work titled *Weather Cranes* (2007). By attaching weather-responsive colored indicator lights to the pulley wires of the cranes, Giddy was able to register local shifts in temperature and humidity. He also connected heat-responsive lights to the crane's cabins, which glowed with intense colors as the temperatures became more extreme. This highlighted the human costs of working in temperatures exceeding 40°C in ways that were immediately clear to people in the local environment.

David Stephenson's photographs of nature intervened in by technology offer a powerful reminder of the ongoing tension between modernization and the environment. His first series was of Lake Pedder in Tasmania (Australia), which was the source of the world's first green collective in 1970, one year before Greenpeace was founded. Born in the United States but migrating to Australia in the 1980s, Stephenson's first works in Australia depicted the famous environmental debates surrounding the Franklin Dam in 1982. Using photography and video art, his work explores the aesthetics of representing the environment. In the series *The Drowned* (2002), Stephenson investigated the human specters left on nature around Lake Pedder. As the catalog for *In the Balance* noted, "The corpses of dead trees stand as silent historical witnesses to the complex changes which technological culture perpetrates on the environment" (MCA 2010). This exploration of the various dimensions of intimacy to and within nature, and its entanglement with technology, is a persistent theme in many artworks in the region, where documentary photography has historically played a powerful role in representing the environment and engaging audiences as activists (Palmer 2014).

Many unforeseen environmental problems have arisen from large-scale national energy projects such as the Three Gorges Dam project in western China. The Three Gorges project has had a massive social impact on the lives of millions of people, and it is these human and social costs that are at the core of most cultural interpretations of the effects of the dam. For example, the Chinese filmmaker Jia Zhangke's award-winning documentary *Still Life* (2006) focused on the widespread impact of the dam by concentrating on how the lives of particular individuals had been changed forever.

In the same year, Zhangke's film *Dong* (East) featured the filmmaker and artist Liu Xiaodong as he painted a group of laborers working on a site near the Three Gorges

Figure 6.4
David Stephenson, *Drowned No. 176* (Lake Pedder, Tasmania), 2002.

Dam, a project that included making his large-scale painting *New Immigrants of the Three Gorges Dam* (2006). This was one of his most persuasive "impressions of reality" (Xiaodong and Tinari 2012), in which the dam's human consequences are apparent in the small, isolated groups of individuals standing in an impoverished landscape in the hills above the artificial reservoir.

Although Xiaodong does not approach narrative with the extended stories Zhangke is able to tell in film, he nonetheless focuses on individualized portraits of the poor, whose anguished faces tell the stories of how the degraded landscape has changed their lives. The profound damage the dam has wrought on the local ecology is also limned in the imagery of the rubble-strewn earth and the forlorn figure of a bird twisted into an unnatural figuration of flight.

The impact of the Three Gorges Dam on the Yangtze River valley is also the subject of a less-documentary approach to film, Chen Qiulin's extensive *River, River* video project. Chen Qiulin comes from the town of Wanxian, which was significantly affected by the Three Gorges project, and although on her website she acknowledges that the dam was designed to produce power in ways that would reduce greenhouse gases, *River, River* memorializes the immense human loss arising from the dam's destruction of local towns and villages, along with their cultural histories.

Chen Qiulin's work draws on the commentary of two contemporary narrators and combines this with the dramatic presence of a troupe of dancers dressed in the traditional costumes of Chinese opera. Rather than responding primarily to the ecological costs of the dam as such, *River, River* offers an elegiac, performative response to the loss of social and cultural traditions as the dance proceeds across the derelict landscapes of abandoned towns before they gradually succumb to the waters of the expanded Yangtze River.

While the works by Jia Zhangke, Liu Xiaodong, and Chen Qiulin respond to the loss of traditional local life in the regional towns destroyed by the construction of the dam, other artists in the region have turned their attention to the imagined communities of its big cities. In Asian cities particularly, a new form of iconography has developed in which the city appears as an image formed in isolation from its broader context. As Williams (2014) has argued, in both its premodern and contemporary manifestations, the city conforms to a binary form of spatial logic, a system of inclusion and exclusion that effectively sustains the separation of urban life from the rural ecologies on which it depends. And while we have acknowledged the ways in which both communications and international capital connect cities in transnational exchange (McQuire 2008), the environmental connections between regional cities and the ways they will be redefined by global climate change remain less visible.

Nevertheless, the recent and growing formations of publics activated by social and mobile media around critical environmental art are now beginning to provide fertile ground for new kinds of environmental activism, such as the online organization of meetings discussed in chapter 3, and a shift toward art as a potential conduit of political agency exemplified by how the WAWA Project responded to the events of 3/11. The intimacy and micropolitics of social and mobile media also have the potential to leap across geopolitical borders. Such media are able to galvanize awareness of how much wider—indeed, *global*—climatic and ecological change will affect the region, and in this sense indicate an unprecedented potential for grassroots environmental activism. These emergent social processes are also complemented by the recent development of regional art platforms for public engagement, as discussed in chapter 5. They provide new social events through which people are beginning to glimpse how significant ecological change is being felt and imagined by people across the region.

Artists' interpretations of the impact of the Three Gorges Dam, like artistic responses to the crises of 3/11 in Japan (see chaps. 1, 4, 7), have sent clear messages about the unforeseen consequences of geoengineering and nuclear power plants across the region and beyond. Many other artists, however, have found new perspectives on the more familiar imagery of the city and its role in shaping both human and natural ecologies.

The Art of Being Urban

The incremental effects of climate change in the Asia-Pacific are often most palpable from an urban perspective. For example, the impact of rapid manufacturing on the environment, especially in countries that are emergent superpowers like China and India, becomes a perceived reality for city dwellers. It is not uncommon for children in Beijing to be kept indoors when the pollution becomes too high. And yet, in contrast to these representations and perceptions of chronic pollution, images of pristine nature also begin to dominate print and online matter. Against this background, we consider the recent development of cultural images of the city as an isolated construction hovering in a wide expanse of undifferentiated space in the work of regional artists such as Cao Fei or Liu Wei from China, and Manabu Ikeda and Enoki Chu from Japan.

The main platform for Cao Fei's virtual image of urban life, *RMB City* (2007–2009), is the abstracted media realm of Second Life, though sections of her floating city can also be seen on video and in selected stills. *RMB City* is digitally configured as an amalgamation of familiar Chinese cultural and urban icons such as the national flag, a giant panda, the Oriental Pearl TV Tower in Shanghai, and the Forbidden City and "Bird's Nest" sports stadium in Shanghai. With her typically low-key sense of irony, Cao Fei has named her city "RMB" in honor of the Chinese currency that she clearly sees as its motivating force. This lightly ironic tone extends to the official political rhetoric of *RMB City* as "the People's City," which, as such, is a site of political as well as economic discourse. This is especially notable in the video *The People's Limbo*, where imaginary figurations of Mao and Marx converse with an avatar from the multinational venture capitalist company Lehman Brothers (which was liquidated in the global financial crisis of 2007–2008).

At first glimpse, *RMB City*, as it appears in the video and playful sound track for *A Second Life City Planning* (2007), appears to be a fairly lighthearted place, though even here air pollution is clearly going to be a problem, as plumes of smoke and fire arise from chimneys and smokestacks. The Three Gorges Dam Reservoir also makes an appearance as "the People's Dam," pouring forth from the gates of the Forbidden City (now known as the Tiananmen Rostrum) to completely flood Tiananmen Square. In the video *Live in RMB City*, however, the artist's tone becomes darker as an avatar known

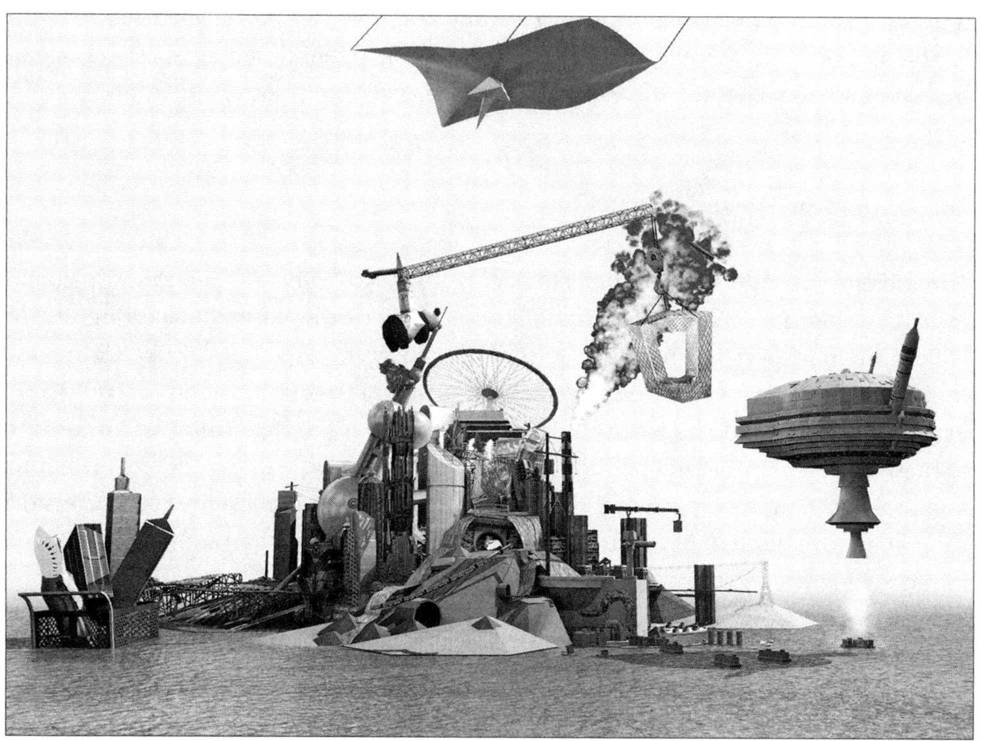

Figure 6.5
Cao Fei (Second Life avatar: China Tracy), *RMB City: A Second Life City Planning Series 04*, 2007.
Digital print. Courtesy of the artist and Vitamin Creative Space.

as Uncle Mars explains to a child born in the virtual world how "the buildings in this
city are merely incarnations of your parents in another time and space; they reverber-
ate with the hollow shells of despair." This is a chilling evocation of life in the small
apartments of countless high-rise buildings in Asian megacities.

Most of the narrative in *RMB City* unfolds from an aerial perspective, as if from a
plane or high-rise building, where the city below appears to combine the optimism of
Chinese urban planning with a sense of fragility and impermanence. The sun is shin-
ing in the abstracted, artificial realm of Second Life, yet the suggestion of how less visible it has become in the polluted cities of the real world is never too far beneath Cao
Fei's created world of appearances.

In some ways, Cao Fei's ironic approach to busy development in *RMB City* is analogous to the work of another young Chinese video artist, Bu Hua, whose playful approach
to urban development in *Savage Growth* (2008) reinforces her critique of the model of

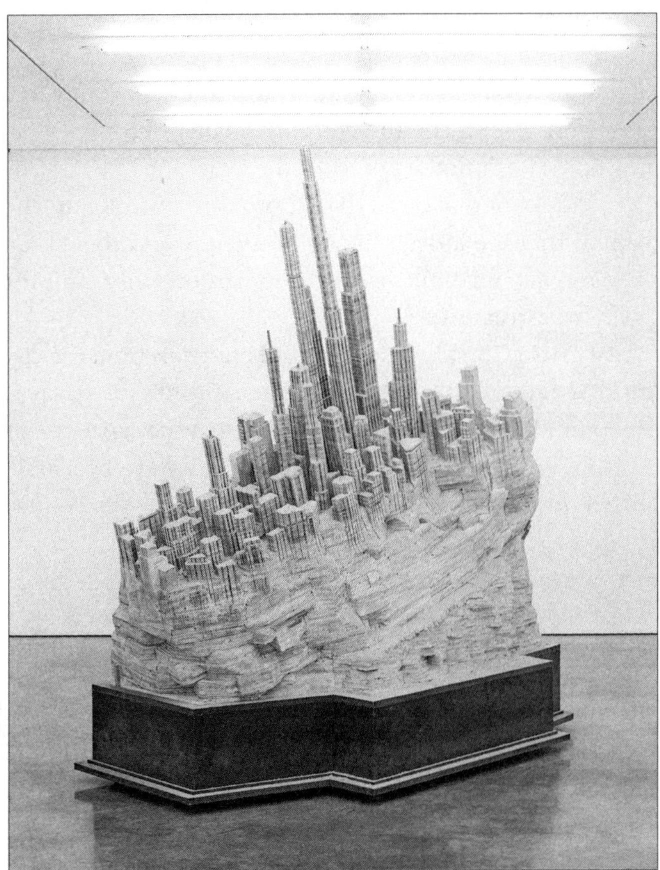

Figure 6.6
Liu Wei, *Inside the White Cube*, White Cube Bermondsey, London, 2012. © 2012 by Liu Wei and White Cube. Photo: Ben Westoby.

perpetual growth. As in *RMB City*, Bu Hua's city has familiar urban icons such as the Oriental Pearl TV Tower in the Pudong district of Shanghai, but it also reaches farther across the region to incorporate an aerial image of Sydney Harbour Bridge. Bu Hua's narrative suggests that unlimited growth will lead inevitably to regional competition, urban decay, and potentially to warfare. Sculptural versions of contemporary cities are also presented from a largely aerial, rather than a more experiential or immersive perspective. For example, the Chinese artist Lui Wei's most recent miniature cities are built up from layers of recycled textbooks, where the surfaces of the works are not refined but rather left as textual signs of recycled materiality.

Lui Wei's miniature cities are seemingly precariously constructed, as if they might slide off their rock bases at any moment. They echo the despair expressed by Cao Fei's character Uncle Mars, especially in Liu Wei's description of the Chinese urban experience as an overwhelming process of development and constant change accelerating to a point where the majority of people "feel numbed most of the time" (Yao 2012). The spatial isolation of the city in Liu Wei's works appears to reinforce its vulnerability and the numbing sense of confinement that the artist refers to. In these works there is an undeniable tension between the implied relations between human ecologies and the self-imposed restrictions of urban infrastructures.

Like Liu Wei, the Japanese artist Enoki Chu also uses recycled materials from the city for his futuristic urban image *RPM-1200* (2008), which is made entirely of salvaged scrap metal. *RPM-1200* is also a miniature, militaristic-looking city designed to be viewed from an aerial perspective, in which all traces of organic life have been subsumed by industrialization. As in the work of Cao Fei and Liu Wei, the defined

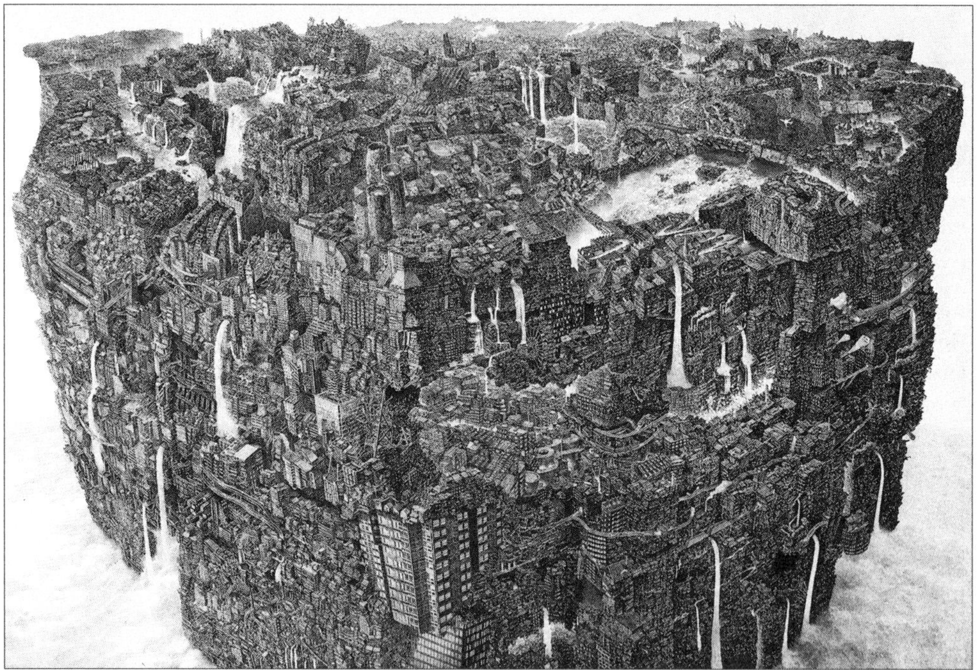

Figure 6.7
Manabu Ikeda, *Ark*, 2005. Collection of Mori Art Museum, Tokyo. © 2005 by Manabu Ikeda, courtesy of Mizuma Art Gallery. Photo by Keizo Kioku.

perimeters of Enoki Chu's city convey a certain feeling of isolation. This contemporary reconfiguration of the miniature, self-contained city as a site of isolation is also clearly a motivation behind the minutely detailed images of urban life in the work of the Japanese artist Manabu Ikeda.

In contrast to Liu Wei and Enoki Chu, however, Ikeda's large-scale drawings are imbued with a sense that although the city appears to be self-contained, it also appears to be fused with a semiorganic structure rendered with more permeable borders. It is, in short, a vision of the city where human and natural ecologies are more closely enmeshed, if uneasily. In Ikeda's work *Ark* (2005), for example, the water flowing from various drains or pipes in the city cascades down to the ocean. The city emerges from the geological bedrock of a mountain plateau. As an image of contemporary life, *Ark* gives the impression that it may well have risen from the ocean itself in the far distant past, which, as an often marginalized evolutionary process, is true of all human endeavors.

Although a miniature plane can be seen making its way into the city—apparently from "outside"—the future of this urban ark appears to reside entirely in the question of whether or not it can continue as a viable structure in its own right, in the center of a surging ocean. The title, *Ark*, suggests that life in the city will be the only means we have to adjust to future change, yet as a complex image combining both the built environment and the organic, it also represents a transitional reconfiguration of relations between the city and regional and global ecologies in which the image of water becomes a vital conduit to the world beyond. In 2013, Ikeda responded to the earthquake and terrible tsunami of 3/11, and the subsequent meltdown of the nuclear plant at Fukushima, with his work *Meltdown*, in which an entire nuclear plant appears to be organically fused to a gigantic glacial boulder as it slides down the curvature of the earth into the ocean below.

In *Meltdown* the giant nuclear complex is caught in suspended motion in contrast with a pristine forest landscape, and it is this moment of suspension, as an accident about to happen, that renders it such a lethal threat to the ocean in which it is about to be immersed. The vulnerability of the ocean to radioactive pollution is clearly a key factor in this work, just as the seepage of radioactive material into groundwater and the Pacific Ocean remains one of the most serious problems at Fukushima (Behrens et al. 2012). Ikeda's title *Meltdown*, however, and the imagery of a city fused with melting ice also convey the idea of a more globally significant event: the potential melting of polar ice caps due to the intensification of carbon emissions produced by big cities. The imagery of water has emerged in the work of a number of regional artists as an effective metaphor for crossing geopolitical borders into new biopolitical territories where

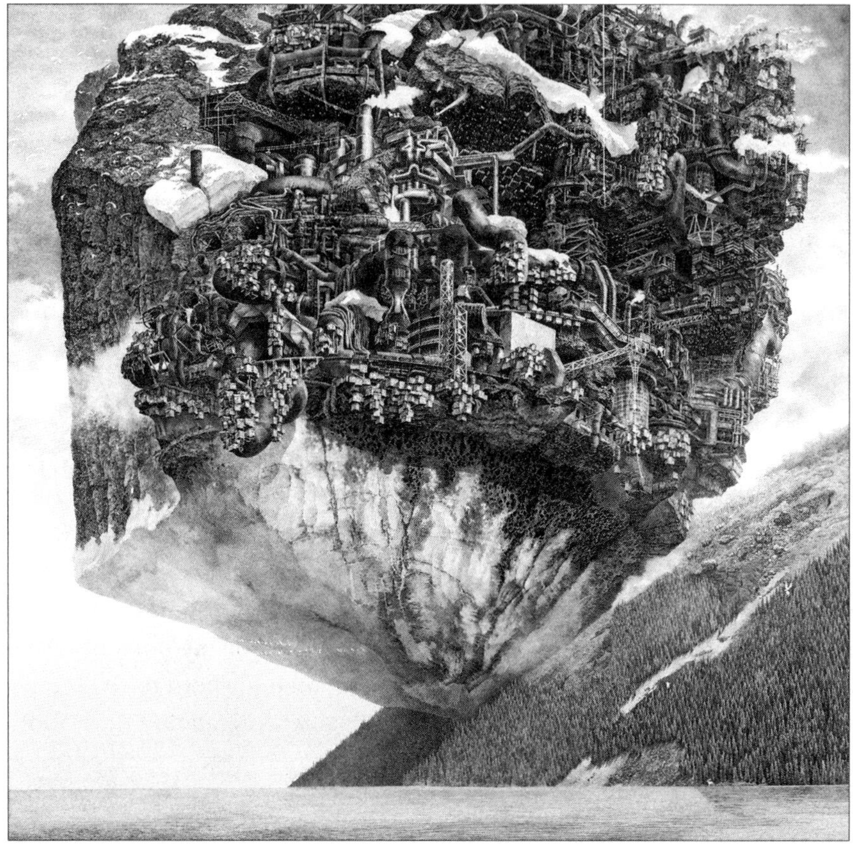

Figure 6.8
Manabu Ikeda, *Meltdown*, 2013. Collection of Chazen Museum of Art. © 2013 by Manabu Ikeda, courtesy of Mizuma Art Gallery. Photo: West Vancouver Museum.

human ecologies cannot be understood as processes developing in isolation from the materiality of nature. This is a nascent tendency, and one that has varied degrees of effectiveness in engaging new publics in the development of ecological problems across the region.

Art and Water: Reflecting Regional Environmental Concerns

Water has become one of the most contested material sources in the region, as well as being a site for environmental pollution and disaster. From the politics of water bottle packaging (Hawkins 2011) and tsunamis to the overheating of oceans predicted by the IPCC (2013), water in its variety of forms plays a key role in many environmental narratives. Bottled water in the region is framed by consumerism and emerging middle classes in the fast-growing Asian cities, where water supply infrastructures do not necessarily keep up with demand in either water quality or availability. At the same time, artists such as Stephen Haley (see chap. 3) demonstrate how the plastic bottles used to package water, and the industries that produce them and distribute bottled water, are complicit in the production of plastic waste that is polluting the oceans.

Because water raises such a set of fundamental and critical environmental issues in and for the Asia-Pacific, it offers an ideal example through which to explore the development of regional arts networks and the ways in which publics are becoming engaged with them. Recent approaches to water through arts practice in the Asia-Pacific indicate how the imagery and tangible presence of water are becoming an important conduit of communication in regional artworks and, as such, express emergent ideas around media and the environment that are relevant across the region.

The Taiwanese artist Vincent J. F. Huang chose a specifically transpacific approach to global climate change in his cultural collaboration with the small Pacific island of Tuvalu when he designed the nation's first pavilion at the 2012 Venice Biennale. Located just south of the Equator, Tuvalu comprises several extremely low-lying coral reef atolls, which are particularly vulnerable to global rises in sea levels (Barnett and Adger 2003). In 2012 Huang took his art projects to Tuvalu itself, where he incinerated an effigy of a mermaid on the shore for his *Dried Mermaid Project*. In his *Floating Bear Project*, an artificial polar bear swam with local children and spent time lying in a hammock in the tropical sun. The children in Huang's video appear to be enjoying his quirky humor, although it is not clear what other locals really thought of the artworks, especially since evidence suggests that many islanders do not want to be seen as victims of global climate change (Mortreux and Barnett 2009).

The people of Tuvalu have demonstrated a high degree of awareness of climate change, but rather than seeing their particular geographical vulnerability as "evidence" in largely Western debates on global environmental crises, they prefer instead to define the terms of their own debate (Farbotko and Lazrus 2012). Huang's real audience seems, in any case, to be international, as his choice of the Venice Biennale suggests, along with events covered by international media such as his *Suicide Penguins*, in which he hung toy penguins by their necks off a bridge in central London. His interactive work for Venice, called *In the Name of Civilization*, was an interactive installation that clearly targeted processes of global consumption by inviting viewers to use a petrol pump connected to a device that slowly decapitated a model of a turtle while simultaneously strangling a model of a polar bear. The lesson was obvious enough to any of the visitors as an example of didactic environmental art (Williams 2014), especially since most, if not all, visitors were required to rely on carbon fuel to get to the exhibition in Venice, including Huang himself. While this in itself does not necessarily weaken the impact of his point, Huang's confrontational approach seems to refute other, more complex, readings.

The decadence of his *The Last Enjoyment before Melting* (2009), however, makes a more nuanced commentary from the position of a Taiwanese artist living in Shanghai to consider how Chinese culture is perceived from the outside. As the local critic Szu-hsien Li (2009) suggests, Huang's use of king and emperor penguins as characters in an artwork inspired by tenth-century Tang dynasty paintings by Gu Hongzhong is based on their introduction to a zoo in Taipei around the new millennium, where they caused a populist sensation. Huang also introduced large-scale sculptures of the penguins, posing as kung fu masters as seen in popular film, in the shallow ornamental pools of the China Europe International Business School in Shanghai. Szu-hsien Li argues that this was as much a commentary on Western perceptions of Chinese culture as it was a reflection on the impact that such a culture may have on climate change: "There is nothing wrong with appropriating Chinese images, but other Chinese may consider it calculated, boring and affected. However, for foreigners it is utterly gratifying to experience the Chinese culture at one glimpse" (Li 2009, n.p.).

Another attempt to traverse borders across the Pacific through art was an exhibition called *Breathing Atolls*. This was curated by the Japan Foundation in the Maldives, which, like Tuvalu, are a group of islands highly vulnerable to any further global rise in sea levels. The project brought a range of contemporary Japanese artworks to the Maldives, such as Fujiko Nakaya's fog and light sculptures, Tetsuro Kano's installations with local live birds, a fish sculpture made of local garbage by Yukihiro Taguchi, and Haruka Kojin's use of various materials in her site-specific installation *Contact Lens*, which drew on locally situated impressions of fluidity and light combined. Of these, Nakaya's ephemeral fog sculptures, which she has shown in various other locations, was the most physically interactive artwork, flowing across the local landscape in ways that immersed both human and natural ecologies in a shifting cloud of water vapor. In this immersive environment, various aspects of the island became, in turn, either visible or invisible so as to subtly reveal the susceptibility of the Maldives to rises in sea levels.

The materiality and experiential aesthetic qualities of *Breathing Atolls* can be evaluated in direct contrast to an online conference held recently in Second Life called "Climate Change: In the Real World There Is No Second Life," in which the Republic of Maldives, the Alliance of Small Island States, and DiploFoundation discussed the threat of global climate change. Like *Breathing Atolls*, the conference made important points about the growing urgency for public engagement with the issues of climate change and ecological deterioration, but the comparison also reveals the capacity of art to engage publics experientially. These two communication platforms are far from mutually exclusive, since works from *Breathing Atolls* are available on YouTube, the stilted abstractions of Second Life have the capacity to engage global publics, and the discursive and experiential differences between the two point to the nascent qualities of both fields.

In an installation for an Australian exhibition, *Heritage* (2013), the Chinese artist Cai Guo-Qiang, like Manabu Ikeda, conveyed the imagery of an ark. In this work, a range of large, artificial, yet highly lifelike animals were arranged to drink from the edges of a massive pool of clear water installed in Brisbane's Gallery of Modern Art. As in the legend of the ark, Cai Guo-Qiang brought predators and prey together peaceably to drink from this pool, and though the naturalism of the artist's rendition of the animals impressed local audiences, in many ways it was the pool itself, and the possibility of its impermanence, that rendered the various species of animals more vulnerable.

The affective and emotional response to the changing role of nature through the intersection of art and environment is explored by the Japanese artist Takashi Kuribayashi. Kuribayashi's approach to his viewers often requires bodily participation in immersive artworks designed so that his reconstructions of nature can only be seen from inside the work itself, and from a ground-up perspective in relation to the landscapes or imagined creatures of his installations. In *Aquarium* (2006), for instance, viewers were required to put their heads, one at a time, into a glass cube in an aquarium tank to experience a re-created oceanic environment. Similarly, in *Wald aus Wald* (Forest from Forest) (2012), viewers entered from "underground" to put their heads through holes in *washi* paper arranged as an imaginary forest floor so that they could experience the effects of the snowy winter landscape that surrounded them.

Although Kuribayashi's works are presented in urban galleries or online videos, they represent subterranean or submarine perspectives rather than the aerial perspectives of the images of the city we considered earlier, and arguably are consequently offering a more immersive experience of imagined ecologies. The public response in

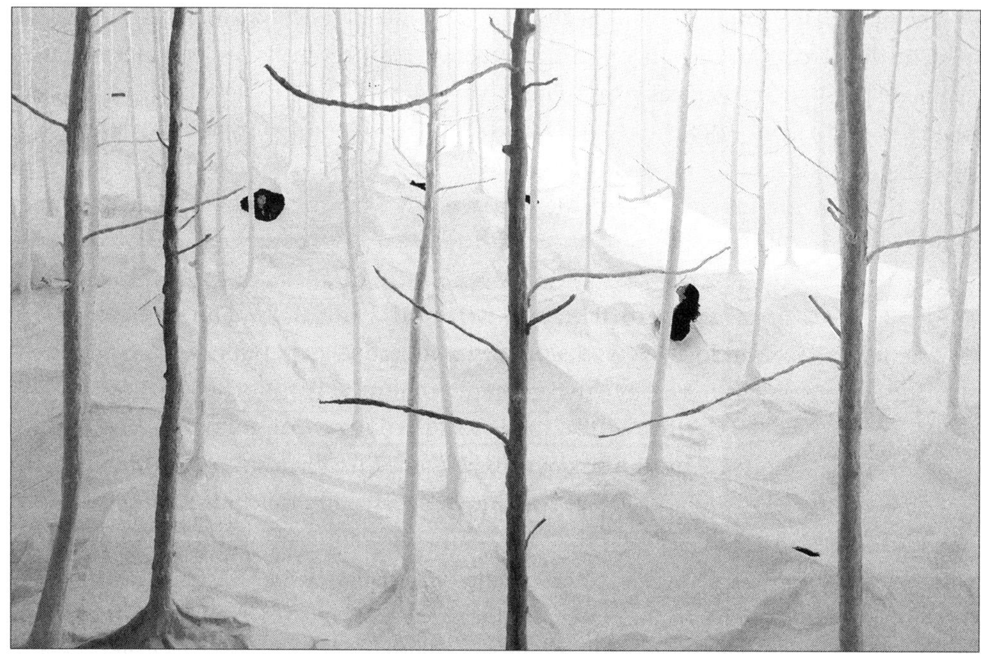

Figure 6.9
Takashi Kuribayashi, *Wald aus Wald*, 2010. Courtesy of the artist and photographer. Photo: Osamu Watanabe.

Japan to the nuclear contamination of Fukushima and the Pacific Ocean after 3/11 is part of a much longer history of Japanese ambivalence toward nuclear energy after the bombing of Hiroshima and Nagasaki by the United States at the end of World War II.

This memory has, for example, been a recurring theme in the work of the Japanese artist Kenji Yanobe, who wore his yellow antiradiation suit into the exclusion zone at Chernobyl in the Ukraine long before he claimed his installation *Antenna of the Earth* (2000) for Williams's *2112: Imagining the Future* exhibition in Melbourne. This exhibition was Yanobe's latest statement about the dangers of the 3/11 nuclear meltdown. The painful memories of the events of 1945 were also made explicit in a work by Cai Guo-Qiang during a visit to Japan, when his *Black Fireworks: Project for Hiroshima* (2008) was shown as part of his long-term project *The Century with Mushroom Clouds*. In this video performance, black fireworks crackled and plumed into a ghostly black tree of smoke, which descended like a mushroom cloud precisely over the ground zero point in Hiroshima.

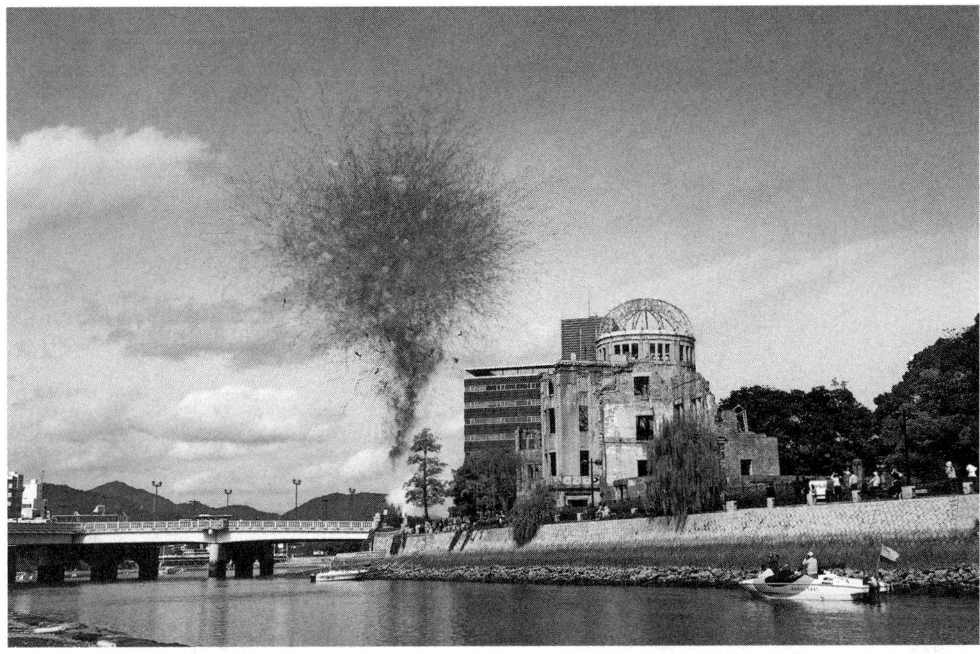

Figure 6.10
Cai Guo-Qiang, *Black Fireworks: Project for Hiroshima*, 2008. Realized at Motomachi Riverside Park near the Atomic Bomb Dome, Hiroshima, October 25, 2008, 1:00 p.m., 60 seconds. Photo by Seiji Toyonaga, courtesy of Hiroshima City Museum of Contemporary Art.

Kuribayashi's response to the later events of 3/11 took his approach to immersive environments much further when in 2012, a year after the nuclear meltdown, he returned to make a video called *Tornado* on the border of the exclusion zone twenty kilometers from the coastal plant at Fukushima. This powerful work combines images of the site with scenes from Kuribayashi's workshop, where he and his team glued contaminated ferns to decorate the surface of a surfboard. Kuribayashi then wore a wetsuit and protective mask and headed out on the surfboard into the contaminated sea at the edge of the exclusion zone.

Interestingly, we see this performance from dual perspectives—one from a camera onshore, another "inside" from a small camera attached to Kuribayashi's head. In this work, it is the artist himself who enters into a completely immersive and potentially toxic space, yet the metonymic quality of the hemispherical ocean as it appears from the lens on his helmet is reminiscent of global space, just as the curvature of the earth in Manabu Ikeda's *Meltdown* suggests ecological decay on a global scale. And although

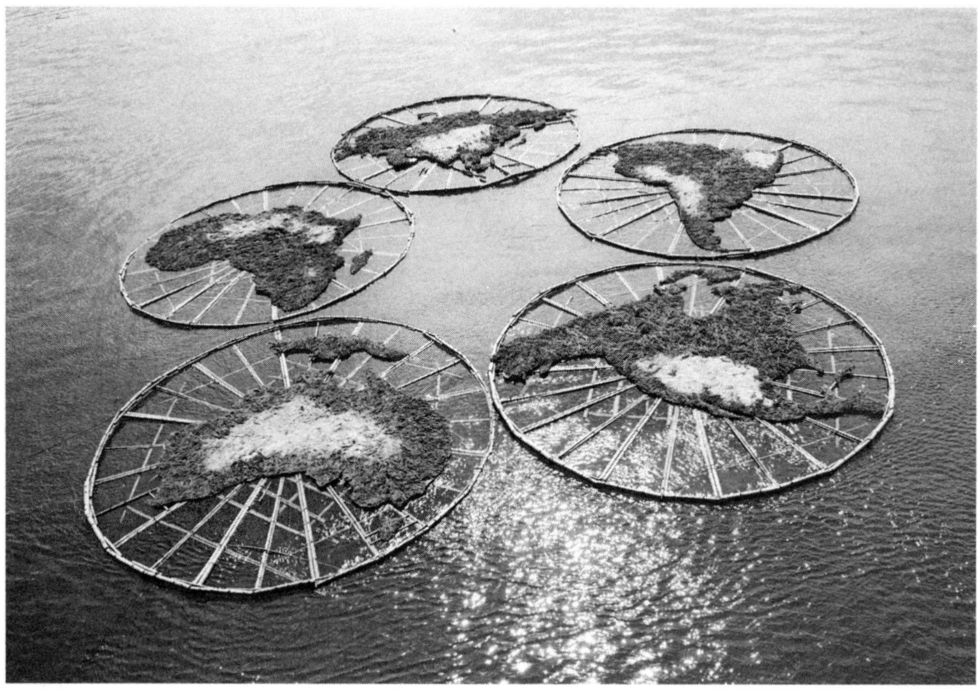

Figure 6.11
Ichi Ikeda, *Five Floating Islands*, 2008. Courtesy of Ichi Ikeda Art Project. Photo: Tatsuro Kodama.

water is much more fundamental to Kuribayashi's work, both artists clearly draw on the affective qualities of water as a channel to reimagining not only local ecologies but also their interconnectedness with regional and global space.

Other works by Kuribayashi, such as *Island* (2010), *Wolkenmeer* (Sea Clouds) (2012), and *Erde* (Earth) (2012), focus more explicitly on extending local land masses into oceanic and global space. In keeping with the theme of water as a medium of transforming spatial boundaries, the Japanese artist Ichi Ikeda has used water as his main medium in ways that draw on its mutable qualities to reimagine our relations with material ecologies. In *Five Floating Islands* (2008), for example, Ikeda created artificial islands floating on the Kedogawa River in Kagoshima, representing the five continents of Eurasia, North and South America, Africa, and Australia.

This work is a reminder of how all biopolitical formations depend on complex global hydrological systems, which, as Ikeda explains on his website, also calls our attention to "the actual changes in the patterns of hydrological cycles produced by climate change" (Ikeda 2008, n.p.). In an interview, Ikeda recalled the public statement he made as a guest of the United Nations when he asked people "to realize that water, source of our lives, is the medium to move beyond borders, everyday customs, histories and cultures, in order to realize the profound exchange between human beings and environment. I call this thought the 'United Waters' an alternative to the United Nations" (Shizimu 2012, n.p.).

Ikeda's concept of what he calls "water thinking" works as a conduit for a more fluid sense of an imagined community that incorporates both human and nonhuman ecologies. Along with other regional artists, he clearly seeks to extend art practices into a more phenomenological relation with nature. From this viewpoint, nature is regarded no longer as something "over there," beyond the boundaries of the city, but rather as processual and intimately enmeshed with the experience of everyday life.

Conclusion: The Role and Potential of Art as Critique in the Context of Climate Change

In this chapter, we have seen how the artists of the Asia-Pacific have reconfigured social practices in which images of nature are reified in cultural commodification, and have begun to reveal both our intimate connections with the materiality of nature and our increasing vulnerability in the context of the anthropogenic causes of climate change.

Much of the arts practice that echoes the ecocritical tradition in scholarship has tended to focus on issues that are generated within nation-states and within particular urban configurations. Nevertheless, as interventions that are being developed from within Asia-Pacific countries and cities, these traditions in arts practice are emerging as a powerful acknowledgment and critique of the problems associated with industrial development, urban intensification, and the concomitant issues of pollution, contamination, and climate change. Taken collectively, they begin to map out, through art, a critical vision and refiguring of the ways in which the region's development is entangled with nature and the environment. They demonstrate issues that are transnational across global, regional, and local flows.

We also focused on projects that have taken a further step toward acknowledging how environmental issues expand across the region and connect different nation-states, cities, and other areas by acknowledging that climate change, disasters, and critiques are not simply located in single sites in the region, but the configurations of things and processes that bring them about are constituted through entanglements that draw together geographically, and also digitally, dispersed flows, objects, and ideologies. Following the critiques produced by such projects, it is easy to imagine how climate change can be viewed as something that is happening not *to* the region but *in* the region and at its interface with other global realities. The lesson from these bodies of work is that we need to take this view from within as our starting point; that is, we need to understand climate change and the ways in which critiques of the processes that bring it about can be incorporated into practical action by understanding its constituents from within, and like the artists whose work we have reviewed in this chapter, engaging with them from within. In the next chapter, we explore the role of collaboration as a means for creative intervention.

7   Emergent Paradigms for Collaboration

In this chapter, we continue the discussion of art and artists in the Asia-Pacific by examining how new models for collaborative cross-cultural art projects have developed in the region. Through the example of how networked and social media are now implicated in reimagining artistic subjectivity and engagement, we argue that artwork emerging from the region is providing new paradigms that are also relevant for understanding collaboration.

Contemporary urban screen culture has undergone a significant shift away from "packaged" forms, such as those disseminated through television and billboards, toward media that enable participatory modes, such as smartphones. As we suggested in chapter 4, the use of smartphones tends to change how people experience the relationship between public and private domains by bringing together the public and private in ways that traverse digital and material environments. In this chapter, we introduce a discussion of how public art might be experienced through the concepts of copresent visualities and "intimate" publics. These ideas are taken up in more depth in the next chapter.

Concepts of mobility and movement also offer us a way of thinking about how relationships are forged between art and publics in the Asia-Pacific. Importantly, the publics generated through the projects discussed in this chapter are mobile, not only in that they are users of mobile technology, but also because they operate in and across localities, regions, and nations. As we have emphasized throughout the book, art is not only a producer of aesthetic meaning but can also generate new social and political meanings. While participatory and collaborative practices are not new in art, they are increasingly prevalent in shaping how contemporary art is encountered, how place is experienced, and the forms of mobility that constitute these.

As we discussed in chapter 5, the use of online platforms has become increasingly popular in media art. In this chapter, we show how this move has also begun to transform the ways in which collaborative art is played out. We argue that the forms of public interaction and collaboration facilitated through contemporary screen cultures, particularly mobile media, can be understood as practices of what Papastergiadis (2012) has called "aesthetic cosmopolitanism," which facilitates new forms of transnational agency and entangled relations of place.

Reimagining Subjectivity through Collaborative Engagement

Historically, collaborative and participatory art practices have been developed across a range of genres over the twentieth and twenty-first centuries. Of most interest to our discussion here is the emergence in the 1990s of art practices focusing on social relations, including Nicholas Bourriaud's (2002) notion of "relational aesthetics," or what Grant Kester (2004) defines as "dialogic practices." These works foreground the temporal, social relations of artworks rather than their material forms and structures. In a contemporary context, this emphasis on arts practice coincides with the emergence of the participatory and collaborative cultures of Web 2.0. It thus creates a wider environment in which we see a coherence between the principles and practices of participatory art and online activities emerging through digital platforms.

Collaborative art projects are commonly celebrated for the ways they challenge deep-rooted ideologies of individual artistic subjectivity and the autonomy of the artist as a "creative genius." Acts of public participation engaged through particular forms of collaboration, which focus on the construction of events as artworks, are also celebrated for the manner in which they break down perceived distinctions between art and everyday life. Such idealizations are difficult to avoid. For example, as Kester describes:

Here the sublation of art and life is sought not through the introduction of quotidian material into the sanctified field of the art object, but through the dismantling of the artistic personality itself in a splay of mediatory practices and exchanges. Deprived of the venerating mantle of the gallery and the museum, the artist is rendered less majestic, but also more accessible, better able to reveal creative insight as a shared human capacity rather than a divine gift. (Kester 2008, 61)

Individuality, autonomy, and agency are certainly transformed through acts of working together. However, if the focus remains on such "dismantling" of individual identity into nonhierarchical relations, then anything other than this defines collaborative engagement as having failed. In this sense, approaches and critiques of collaboration tend to focus on measuring its value and meaning in relation to ideas of authenticity—how the processes, activities, and outcomes produce an equality of relations, or if they reinscribe authorial status on particular individuals (usually the artist), and how they mark insider–outsider positions or erase them. For example, Hal Foster (1994) gives a scathing critique of the popular practice of international cross-cultural collaborative art projects, which offer all the signs of ethnographic approaches to site-based forms of production but in fact reinforce the advantaged position of the artist (against that of local participants):

Consider this scenario, a caricature, I admit. An artist is contacted by a curator about a site-specific work. He or she is flown into town in order to engage the community targeted for collaboration by the institution. However, there is little time or money for much interaction with the community (which tends to be constructed as readymade for representation). (Foster 1994, 17)

According to Foster, such projects become a form of "ethnographic self-fashioning," lacking any substantive self-reflection and critique (1994, 17). Foster's critique is echoed by other critics of the types of collaborative projects that are celebrated through international biennial mega-events, which are seen as aestheticizing new forms of capitalist exploitation (Martin 2007) and providing a "guilt-easing" salve for the art world without engendering change for participants or local communities (Kester 2011).

Such criticisms shift the focus toward the ethical concerns of collaborative actions. In a series of well-known and heated dialogues, Kester and Claire Bishop debated the value of aesthetic and ethical concerns in evaluating participatory and socially engaged art practices. For example, Kester (2004, 2011) focuses on examples of non-hierarchical participatory relations to emphasize the ethical relations of collaboration. However, Bishop (2006a, 2006b) challenges the overemphasis on "good intentions" and "consensus" as the site of political agency of such collaborative art practices and wants to reinvest them with antagonized relations. Bishop aims to situate socially engaged forms of art in the history of avant-garde art and therefore advocates for the autonomy of art (through aesthetics) in order for it to remain politically critical. In doing so, Bishop reinstates boundaries demarcating art, trying to "rescue" art's blurry overlap with social activism (Kenning 2009). This is a delicate dance between aestheticizing collaboration and emptying it of meaning, as Foster observes, and limiting the social and political agency of collaboration to ethical relations. This line is further complicated by the co-optation of art by political activism (Raunig 2007).

To avoid getting stuck in binary positions between aesthetics versus ethics, or demarcating art from social activism in ways that can be construed as artificial, this chapter critically examines how collaboration generates new forms of subjectivities, particularly in a transregional context and using mobile technologies. Collaboration emphasizes subjectivity as mobile and always in formation—never residing *in* the thing, but formed between relations (Deleuze and Guattari 1987; Derrida 1982). As such, while it is important to reflexively consider the relations of power in collaborative projects, these should be approached not as fixed and static but as open to constant transformation through acts of participation and exchange. As Kester argues, artists occupy multiple positions working with communities; there is "a toggling back and forth between inside and outside, engagement and observation, immersion and reflective distance" (2011, 90) at any given time in a project. The forms of copresence generated through mobile screen cultures, discussed in chapter 4, also problematize fixed and static ideas of identity.

To demonstrate some of these complex relations of identity, we discuss two approaches to collaboration from the region. Both of these art projects examine human impacts on the urban environment: *Stereopublic: Crowdsourcing the Quiet* is an online participatory project using smartphone applications and geotagging to collect sounds of "quiet" urban spaces, and *The Library* by Soundpocket (Hong Kong) is an online library of sounds created by individuals using whatever recording technology is available to them and posted to an online "sound map." Soundpocket is a Hong Kong arts organization established by Yang Yeung to promote sound art and listening practices.

Stereopublic (http://www.stereopublic.net) is directed by the Australian sound artist Jason Sweeney and gained prominence through its recognition by the TEDCity2.0 prize. The work offers a new approach to acoustic ecology, which considers urban noise as the pathology of "sick" cities (influenced by R. Murray Schafer and the World Soundscape Project started in the late 1960s, which critiques the noise pollution of cities). Sweeney is promoting "sonic health," and the project is designed to create "earwitnesses" who record urban quiet spots on their mobile phones and upload the recordings to the website. The recordings are then recomposed by Sweeney and uploaded to an interactive global sound map. The project emphasizes mass participation achieved through collecting sounds from urban sites around the world and creating an interactive online map.

However, in what Sweeney refers to as "collaborative remix," he recomposes the sound into artworks, thereby reinstating the artist as specialized auteur (on the website, he is credited as "director"). Thus, in this work, the individual artist remains at the core of the project, and the participants are a form of artistic outsourcing. It is the artist who controls the finished works. In this instance, "participatory" does not equate to horizontal relations of collaboration. Like the concept of platforms discussed in chapter 5, the rhetoric of collaboration can hide the individual, specialized position of the artist as primary author in relation to the mass of participants, who source and select the sounds through local fieldwork.

In contrast, *The Library* by Soundpocket (http://www.thelibrarybysoundpocket.org. hk) is more focused on creating an open database of urban soundscapes with less focus on editing or curatorial direction. There is some "top-down" input through the placement of the work, the visual aesthetic of the online platform, and the conceptual framing and structure of the library, which are managed by Soundpocket. One consequence of the open approach to participation is that the quality of the works is inconsistent, but it is arguably a more egalitarian approach to participation rather than hierarchically positioning the artist as author of the "art" content. *The Library*'s approach requires no such distinction between users.

What is innovative about both projects is the use of mobile and social media to engage with a range of publics and their relationship to site. The works themselves require physical audio recording of places, identified using the GPS capability of smartphones; however, these recordings can be downloaded and listened to in any number of locations around the world. This has the potential to set up dynamic and complex interpenetrations of place offline–online and local–global, which generate new types of social relations within mobile and intimate publics. While in one sense the recordings represent specific places, how they are accessed—from anywhere at anytime and replayed in any number of new and different contexts through mobile technologies and social media—causes them to become part of *other* localities and experience of place. It is in this way that a global sense of the environment becomes part of local experience, art becomes part of the world, and local political meanings can be overlaid on the work. This makes these works an innovative example of contemporary screen cultures and art.

Defining participation and collaboration in art projects in the context of contemporary media culture inevitably engages with discourses of the participatory and connected cultures of Web 2.0. The rise of online networked media has marked a "participatory turn" (Burgess and Green 2009) that centralizes "vernacular creativity" (Burgess 2007). These features are exemplified in both *Stereopublic* and *The Library*, which explore the idea of the user as a hybrid producer or "produser" (Bruns 2008). The participatory and collaborative possibilities afforded by digital and social media are already engaged extensively in digital political activism globally, and they have been used to mobilize public environmental critique, such as in the nuclear power protests in Japan after 3/11. This contemporary participatory culture has often been discussed in terms of the rise of UCC and modes of distribution, which have the potential to challenge conventional relationships between consumers and media producers of online media (Jenkins 2006).

Yet, as Burgess and Green point out, cultures of participatory media such as You-Tube are "as disruptive and uncomfortable as they might be potentially liberating" (2009, 10). Tensions are situated around issues of access, representation (who gets to speak, what they can say, and what gets attention), authority, and specialized knowledge. Burgess and Green (2009, 57) deliberately use the term "participants" to refer to all users of YouTube—private individuals, businesses, or organizations—because the content that is viewed, commented on, used, and circulated via YouTube has value and meaning based on its genre, its uses within the site, and its relevance to the everyday life of other users, rather than being based on a differentiation between producers of content (e.g., mainstream media companies or amateur videobloggers). This is an important model because the discourse of participation is often based on differentiating the political agency and power relationships of individuals, or the public collective, against corporate, state, or institutional interests. It also extends the understanding of participants as not only the content creators but also the audiences, who view, comment on, share, and rate "as practices of participation" (Burgess and Green 2009, 57).

Rethinking relations within the collaborative process in this way allows us to consider them as entangled and transversal and for them to be as much about the relations between participants as between the participants and place—as a meshwork of relations. Rather than idealizing collaboration or reducing it to being collusive with market forces, we need to "evaluate the capacity of collaborative art to redefine its aesthetic materiality in the way it 'traverses' the subjectivity of diverse groups of people" (Papastergiadis 2011, 277). The focus then shifts away from strict binary models of artist and participant toward reexamining the complex relations between artists, artworks, publics, and place. These subjectivities are clustered around political issues of environmental concerns and can be understood as "transversal" relations (Papastergiadis 2012) or what Gerald Raunig (2007) refers to as "transversal concatenation." The artwork mediates the act of participation and the self-awareness of participants in this dynamic. This is where the sociopolitical agency of art and collaboration is evident.

Models of Collaboration in Art

While in practice collaborative art takes many forms, one of the dominant models discussed has been collaboration between artists (Crawford 2008; Green 2001). Yet from the art collectives of the 1960s (Gutai in Japan) and 1980s (the Stars Group, China) to more recent examples of mass participatory and collaborative practices (Hans Ulrich Obrist's "Do It" exhibitions) and more modest and localized examples (Hong Kong's

Soundpocket), there have been a myriad of approaches to collaboration beyond the limited focus of artist-to-artist engagement. With the post-1990s focus on relational aesthetics (Bourriaud 2002) and socially engaged practices of participatory art (Bishop 2006a, 2012; Kester 2004), the concept of collaboration has been applied to broader notions of public participation, both within and outside conventional institutional settings such as galleries (Crawford 2008; Downey 2009; Raunig 2007; Stimson and Sholette 2007). However, many of the relational practices identified by figures such as Bourriaud, while focused on social engagement and participation, actually remain constrained inside the social spaces of the art world, rather than extending public engagement beyond these sites (Frieling et al. 2008).

More recently, Kester (2011) has identified new forms of collaborative practice emerging through global interconnections, which are resulting in new forms of artistic identity based on "reciprocal creative labour" rather than conventional ideas of artistic autonomy or localized communities. These new forms involve not just working with other artists or general participants but working with urban planners, NGOs, and communities. These art practices challenge the ideologies of neoliberalism and globalization, including institutions of mainstream art, and advocate for long-term involvement and interactions between artists and local communities, with an emphasis on producing solutions to key sociopolitical issues through collective actions. The shift toward project-based transnational forms of collaboration and models of coproduction distinct from fixed groupings and long-term collectives of previous generations has been identified by Papastergiadis (2012) as a new form of "vernacular cosmopolitanism."

As Kester identifies, these new forms of transnational collaborative coproduction are not anti-institutional but rather reconfigure partnerships to challenge binary models of individual or public interests versus corporate, institutional, and government interests. They operate outside the so-called biennial model and offer new local and regional perspectives that are consciously cosmopolitan. In contrast to Kester's overemphasis on convivial relations, and Foster's criticism of superficial fly in/out models, these projects actively partake in more reflexive forms of engagement. They do not always aim for smooth affinities, and they offer more complex sets of relations than demarcated insider–outsider positions.

While this idea of reflexive coproduction achieves a model that acknowledges all participants (including corporate partners, etc.) along the lines of the YouTube model, how these relations are experienced never quite escapes the thorny and uncomfortable issues of hierarchy, labor and power relations. Like the use of "hybridity," which attempts to open up and acknowledge multiplicity but so often ends up smoothing over difference into another third space (Maharaj 1994; Mosquera 2003), the evaluation of collaboration often focuses on how successful it is at producing even, smooth relations of power. In contrast, one can argue that in the actual ruptures of collaboration—what Ien Ang (2003) defines as the "incommensurability" of cultural translation—the fissures between participants are often the most compelling and productive sites of collaboration. It is these frictions that compel further dialogue and avoid projects becoming short-term, self-focused endgames (Sharp 2010).

In this model, the artist is not outside or detached from participants but is part of a copresent place, shaping the process of meaning making. Likewise, the participants (who can, of course, be artists) recognize their own agency and how they interact with others in knowledge production, not just as a step toward "rapport" but also as "epistemic partners" (Papastergiadis 2012, 173). Papastergiadis looks to Rancière (2009) to engage an emancipatory political potential in art based on social relations and ideas of "equality of intelligences" between collaborators. Critically, the key outcome of equality is not achieved through universalizing connections but rather amplifies art's role in mediating new social possibilities. These new social possibilities do not just involve the transmission of ideas but also rather actively construct the place where these ideas are communicated. The transnational art project *Spatial Dialogues: Public Art and Climate Change*, discussed in chapter 4, is demonstrative of some of these dynamics.

Spatial Dialogues

Spatial Dialogues provides examples of the types of transnational collaborative practices that engage intimate publics. This three-year project developed a number of platforms (or outcomes) with different artists participating at different times. Collaborative projects between artists occurred in Melbourne, Tokyo, and Shanghai. The project was facilitated through online media (Skype, Twitter, Facebook, and e-mail) and site-based artwork (performances, exhibitions, and workshops). It is indicative of the type of transnational project focused on art as both aesthetic production and part of networks of social knowledge production—in this instance, environmental issues of climate change and new forms of public engagement.

While a significant portion of the project's research focused on generating artworks—online and offline—the processes of collaborative engagement are as much an outcome as the artwork itself. As such, *Spatial Dialogues* recognized the diverse methods being deployed in collaborative art to renegotiate the social and political agency of art, particularly regarding environmental change. These projects can be understood not only through the acts of collaboration but also through the methods used to document and represent them. They foreground the role of art not only in representing the environment but also in actively forming place, producing sociopolitical meaning, and generating publics. This social context for art production—which includes preparation, curation, and fieldwork—becomes part of the artwork. These "vernacular" methods of collaboration, which are a part of, rather than separate from, everyday activities, do not just embed art into the social; they also become social praxis (Papastergiadis 2012, 181).

Because there is always the danger that there is no recognizable "art" moment in such everyday activities or that, like art and activism, aesthetic forms slip into instrumentalizing, Papastergiadis advocates for some kind of distinction between the act of artistic creation and everyday social activities. For him this reveals itself somewhere along the process as a series of aesthetic decisions (which may or may not be in conflict with the sociopolitical decisions of a project). However, who determines the aesthetic here is not clear. As a project, *Spatial Dialogues* approached collaboration as a

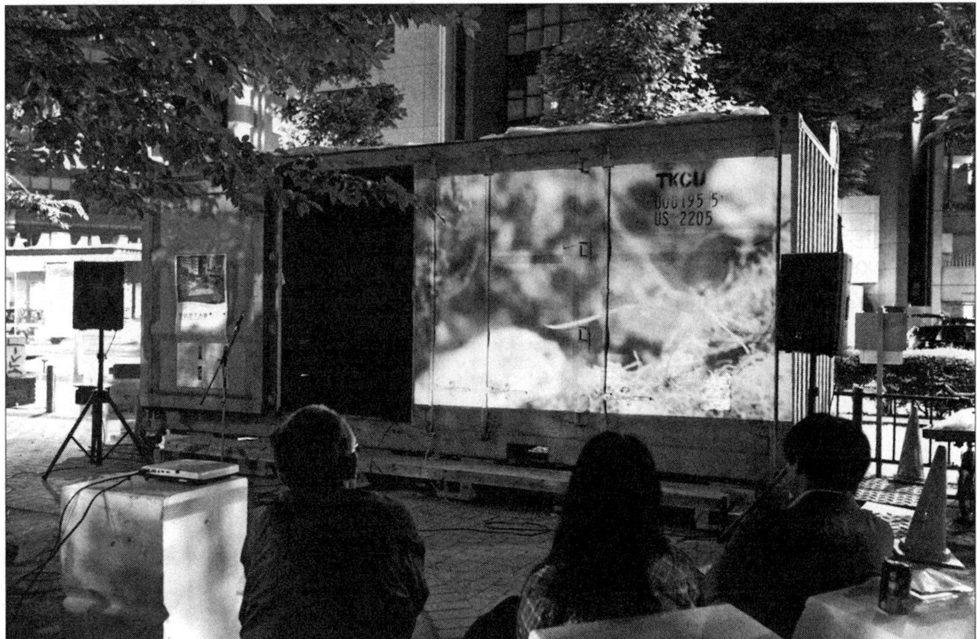

Figure 7.1
Christophe Charles, Dominic Redfern, and Philip Samartzis, *Hidden River*, sound and video performance, 2013. Part of the *Spatial Dialogues* project. Courtesy of the artists.

reflexive and reciprocal process with a range of participants: artists and academic researchers (from Australia, Japan, and China), private industry (Grocon and Fairfax Media), academic institutions (RMIT University, Shanghai University, Musashino University), organizations (the BOAT PEOPLE Association), and online and offline publics. A significant proportion of the project involved working across geocultural spaces (Australia, China, and Japan), including mobile screen cultures. Rather than just focusing on one location, the project worked across sites in Melbourne, Shanghai, and Tokyo and through online social media, such as Facebook, Twitter, and online gaming formats, to engage with mobile publics across these places.

In *Shibuya: Underground Streams* (2013), a *Spatial Dialogues*/BOAT PEOPLE Association collaborative event in Tokyo, artists worked with public participants to generate a series of dialogues and reflections about the changing urbanscape of Tokyo, specifically the Shibuya River, which had been redirected and channeled beneath the concrete roadways of Tokyo. This urban erasure of what was once a significant conduit for local identity and productivity changed the meaning of place and identity, not only for residents but also for visitors. One of the aims of *Underground Streams* was to acknowledge how urban transformation affects relationships between humans, the physical environment, and the changing ecologies of rivers.

Underground Streams produced a number of different works for the project, ranging from art performances, public forums, and mobile gaming to social community activities (riverscape walks and lantern making) and video and sound documentation. The works were often vernacular in nature and aimed to engage the everyday encounter as part of the discussion of the changing ecologies of place, though the project itself created an extraordinary spectacle (based around a shipping container that functioned as site office, meeting point, performance venue, and exhibition space, situated in a park adjacent to one of the major retail strips, Meiji dori, in Shibuya). The works were largely seen as conversations in process, unfolding discussions, opening portals for public discourse online and offline. The "public" here was not an amorphous general mass but rather included local residents, visiting consumers to the shopping mecca of Shibuya, and international participants who could engage with the project via Twitter and Facebook. These are diverse and intimate publics with personal and shared concerns regarding regional and global environmental problems, coalescing as a temporary public through contemporary art.

Rather than uncovering an essentialized "truth" about the identity of the river, the project generated a plurality of perspectives, from locals reflecting on their historic encounters with the river, to others surprised at discovering that a river even existed beneath the road, to others reimagining place through dance, food, mobile gaming, and live performance (see http://spatialdialogues.net/tokyo/shibuya).

Places defy closed interpretations or representations because they are constantly being reformed through social, material, and symbolic processes and practices. This does not deny the specificity of place but acknowledges that place is formed in relation to a range of forces, including its reconfiguring through participatory projects and the subjective encounters of intimate publics. Likewise, the collaborative artworks produced, such as *Hidden River*, were always emphasized as interpretative narratives and subjective encounters that reimagined place rather than seeking to find a singular identity for it. *Hidden River*, while based on video and sound recordings of site, did not seek to establish a direct translation of place. Rather, each artist took his or her own experience and interpretation of the sites and translated them through art, cultural and environmental discourses, and material and social dimensions. In doing so, they reframed them through the collaborative encounter.

This is not to overemphasize collaboration; in *Spatial Dialogues*, many of the artists retained and produced work as individuals at various stages of the project, rather than

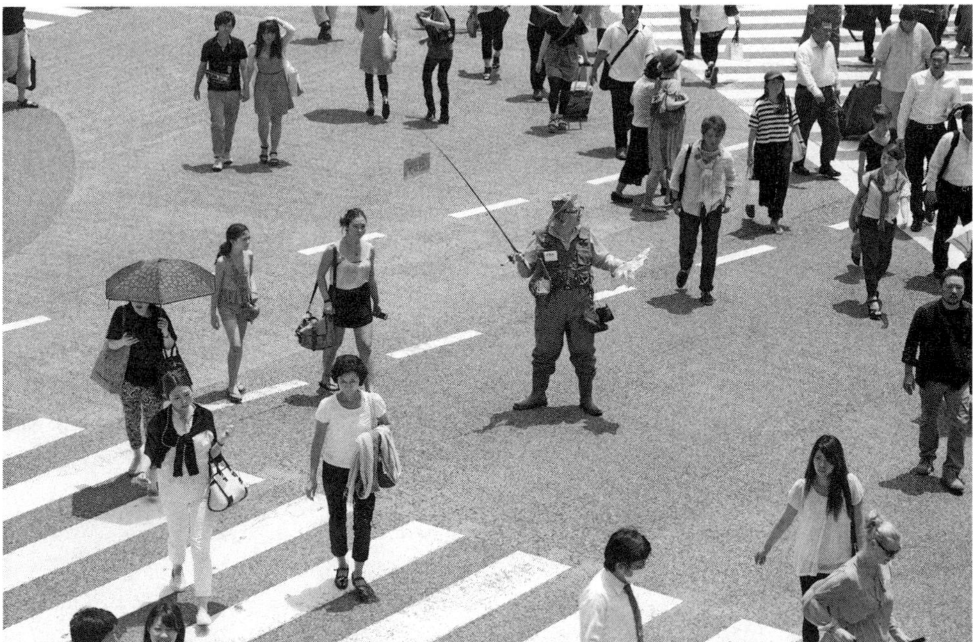

Figure 7.2
Simon Perry, *Lost Fisherman*, 2013. Courtesy of the artist.

in a strict sense of working together in the production of an artwork. Like Kester's toggling, they moved in and out of collaborative roles. Likewise, the public was at times, reassigned from being a participant to being an "audience." If we approach the project, however, as part of a much broader conversation about climate change and examine how a range of participants, including various publics, interacted with the project, then the salient point of the project concerns how these conversations unfolded and influenced the formation of the works. *Spatial Dialogues* was about creating structures within which collaborative relations could occur, rather than imposing a condition of artificially forced methods of individuals working together or expectations of consensus.

One of the project's unexpected outcomes was how differences were amplified through the process of collaborative conversation. The difficulties of working across a range of sites, with a range of people with different background experiences and expectations, created something the artists could push against and through. As one of the artists observed, "It would be too safe if everyone thought alike … It is the counterpoints which create richer works" (personal interview 1, 2011). For other artists, it was precisely the risk and encounter with differences that created a certain level of "freedom" for them to work in new ways and explore new options where they were not bound by expectations of their own (usual) practice (personal interview 3, 2011). This "pushing against" created a productive site (or place) focused on the present: "[Collaboration] allows a focus on the moment. … Art is always a process, and [you] cannot articulate it prior to it being. … Collaboration involves risk. … It is a negotiation which allows different things to emerge" (personal interview 2, 2011).

It is precisely this experience of collision with cultural translations of place that influenced the public performance of *Lost Fisherman* (part of *Underground Streams*). This work depicted an eco-tourist out of place and out of time, searching for a "lost" urban aquatic environment. While the project itself was initially generated out of Australia, and this was the source of its private and government research funding (part of the power relations of labor and environmental discourse that shaped the project), it was localized networks (through the BOAT PEOPLE Association, Musashino Art University, and Shanghai University) that were critical to actualizing the transregional dialogue on environment and screen cultures across the region.

In Shanghai in 2013, the *Spatial Dialogues* writer Linda Williams collaborated with local partner Ling Min in curating the Water and Reflections exhibition and public seminar at the Shanghai Himalayas Museum. This pair of events included Dominic Redfern's video and interactive videos by the Chinese artist Hu Jieming. It was also the platform for an international dialogue between the sound artist Philip Samartzis and the Chinese new media artist Wang Zhang, in which illuminated objects in Shanghai moved in response to Samartzis's recordings of water in the three regional cities. These gallery works were also combined with various public art projects such as floating islands, ice sculptures, and a traditional Chinese landscape made entirely of discarded plastic bags. Another iteration of the project in 2012, *Drowned Worlds*, used the site of Melbourne's Yarra River, where Redfern, Perry, and Samartzis made video, sound, and sandbag sculptures to explore fears about the potential future sea level rise leading to urban flooding.

The process of interaction and collaboration in *Spatial Dialogues* was multifarious, emerging across multiple sites, times, and platforms. It did not aim for a collective or standardized approach to the environment but allowed a diverse range of voices to be articulated in relation to global climate change. The project's length and the opportunities for engagement across a range of platforms provided a way to avoid certain problems identified by Foster, particularly the fly in/out model of collaboration and criticisms of collaboration being used as an aesthetic playfield. Instead *Spatial Dialogues* allowed relationships to emerge in a range of ways: from strengthening long-term relationships between artists, and artists and institutions, to actively forming new networks and sets of relationships, including those of the public. These publics were local, online, and mobile, offering a different approach to considerations of collaboration that remain strictly bound to place as a singular geographic site.

Behind the Scenes: Methods for Negotiating Collaborative Practices

The presentation format of many collaborative projects continues to rely on the exhibition—either a gallery or online space as a form of exhibition—presenting the "finished" works. Such approaches reveal little evidence of the act of collaboration or participation leading to the production of artworks, other than a byline tag to the project being "collaborative." Alternative approaches to this model include projects from the BOAT PEOPLE Association and 3331 Arts Chiyoda's WAWA Project (discussed in chap. 8) that focus on the generation of discourse and art's role in social engagement rather than artworks per se. In these cases, social engagement is considered to be the artistic production.

How we present, discuss, and evaluate collaborative approaches requires us to reconsider production, because this is where fissures of creation are encountered. A "behind-the-scenes" approach has the potential to allow the dialogues and negotiations of collaboration to be considered as much of a "work" as any conventional art object produced, whether visual or temporal (Ashford et al. 2006). Such an approach also allows access to the range of informal negotiations that occur after the formalities are over, nurturing trust between participants, which is fundamental to enabling collaborative processes (personal interview 1, 2011). The postmeeting "cab ride confessions" are where dialogues about place, environment, and personal responses to climate change occur. These are critical to understanding how art collaborates with place in its material, social, and symbolic aspects. Such conversations do not just happen face-to-face but also occur through screens: Skype conversations, SMS, and Facebook, allowing participants to engage with the project in a myriad of ways.

Ethnographic models can be deployed to capture these activities and responses, including interviews, participant observation, and multimodal documentation (video, photography, and reflexive writing). They can also be used to activate all participants as coproducers, including a collaborative dynamic between researchers and the subjects of study: "The extemporaneous and participatory nature of these projects requires the historian or critic to employ techniques (field research, participant-observation, interviews, etc.) more typically associated with the social sciences" (Kester 2011, 10).

As Papastergiadis argues, a collaborative dynamic creates a new role for the critic, theorist, and historian—that of "being there" as part of the narratives of place and participation. This recognizes the role of documentation, interpretation, and the context of encounters in art projects as a form of "visceral cosmopolitanism" (2012, 189), and these encounters form an "assemblage" (185) of activities (or "meshwork," to use Ingold's term). It also shifts the focus away from the finished artwork being considered the primary outcome of a collaborative project, and removes the burden of the art object providing material evidence of collaborative relationships, toward understanding art's "*processes* as integral to the content of the work" (Kester 2011, 9). While "being there" as part of the project might not actually produce a view that is more authentic or intimate, it can offer another useful method of engaging with the project and with place. In *Spatial Dialogues*, embedded approaches were used to understand the context and environments in which the artists were conducting fieldwork. Methods included participant observation, interviews, and photo documentation (see, e.g., the *Spatial Dialogues* website). This provided embodied understandings of place, including the social relations that formed through the project, as well as responses to the geographic site, cultural histories, and experiences of climate change.

As we discussed in the book's introduction, places can be understood as a kind of event that enables forms of copresence, and can be what Massey (2005) calls "stories so far," in that they are always in progress. Employing ethnographic research methods within collaborative projects can be important because such methods enable researchers to engage with the embodied experience of collaborative practice—the "stories" of place and the way place is formed through, and not just represented by, art. Like the ethical issues facing socially engaged practices of art, one of the key challenges in using ethnographic research in art projects is that art is reduced to the communication of an "other" experience where the act of research takes place. When photo-documentation is used to analyze artworks, there is the potential danger that ethical processes are given priority over aesthetic or conceptual concerns. However, if the camera is understood not just as an invisible tool of documentation but as part of the process of mediating relations of place, then fieldwork images become not just about the collection of data (documents of site and collaboration) but part of the process, part of the material and aesthetic stories of place, including that of collaborative relations. As we have emphasized throughout the book, art is not just a representation of place; it is part of the meaning of how place and, in this instance, environmental issues are understood.

A concluding example from *Spatial Dialogues* demonstrates this. During research undertaken in the Saitama Underground Water Discharge Channel (Shutoken Gaikaku

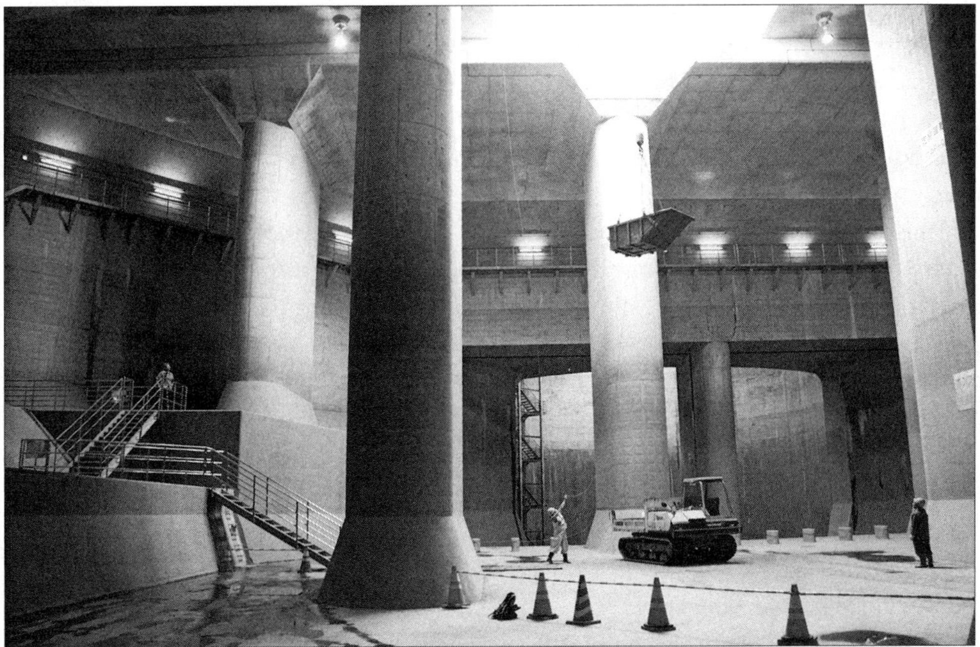

Figure 7.3
Saitama Underground Water Discharge Channel (Shutoken Gaikaku Hōsuiro), 2012. Photo: Kristen Sharp/Simon Perry.

Hōsuiro), it was the sensorial experience of "being there"—the material and psychological engagement with site—that shaped discussions and responses to a post-Fukushima environment and the relationships between increasingly extreme weather events and climate change.

Saitama is a huge underground megacomplex designed to prevent overflow and flooding during heavy rains and typhoons by storing water (http://www.ktr.mlit.go.jp/edogawa/gaikaku/). Fieldwork was conducted over two days in the large concrete surge tank, 24.5 m x 177 m x 78 m, with fifty-nine pillars, which is constructed to hold and then pump water back into the Edogawa River system. Built to withstand and contain forces of nature to protect the city, the overwhelming scale of this underground world appears to be designed to make you aware of the terrifying natural forces, water, flooding, and rain. This was a place hidden underground, and the dank air and moist conditions heightened the artists' awareness of the dangers of stagnant, pooling water and radioactivity. The artists were acutely aware of touching the walls, breathing the air, and making contact with the puddles on the floor. After spending hours underground, it was a relief to climb the hundred steps and surface into the sunny air.

During the time underground, the tanks were being cleaned of silt. The noise of heavy machinery created a constant foundation against which the sounds of artistic work (clicking of cameras, live musical performances, discussions) were activated. The documentation (the sound of the camera) became part of the artistic responses (such as sound recordings of the site). There was a clear sense that being there was not just about being on-site but about being part of a collaborative research project. These interventions (emotional responses, social interactions, and the material presence of equipment and bodies) were equally important and part of the experience of the site.

The artworks produced, whether collaborative or individual, were formed through social engagements and physical encounters with the place. These experiences then became part of a broader imagining of the environment, which had been critically formed and mediated through screen cultures. For example, *Weatherman* (2012), a collaborative multimedia performance work, emphasized the emotional and bodily effects of the events of 3/11 and their ongoing impacts on social consciousness by parodying the kind of television weather presentations discussed in chapter 4. The work presented a chaotic and emotionally charged form of live performance involving video and sound, which engaged the audience in the breakdown of communication and sense of order caused by the tsunami and nuclear power plant disaster. As a

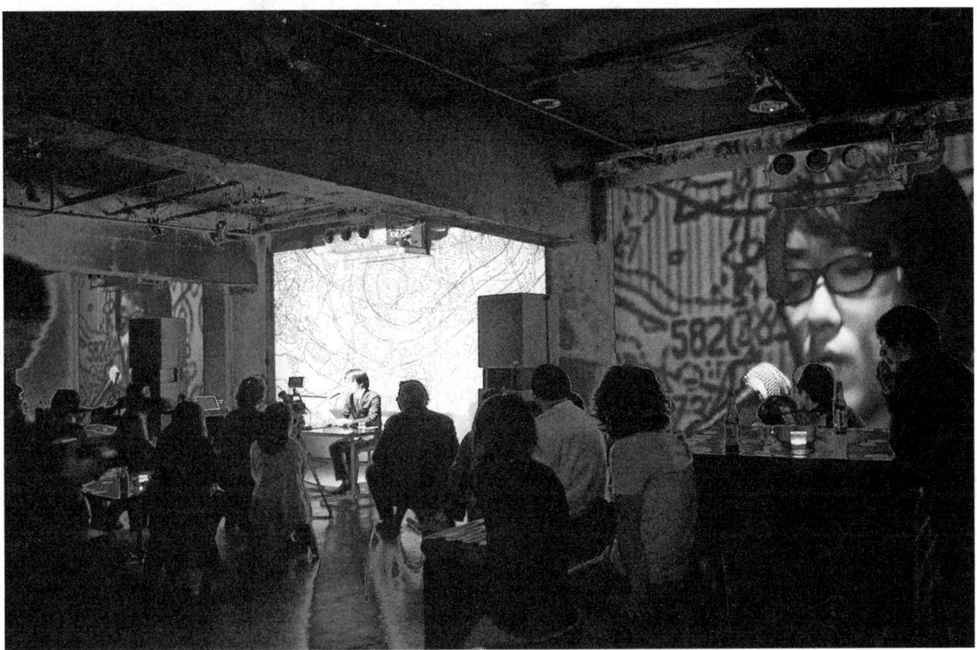

Figure 7.4
Simon Perry, Dominic Redfern, Christophe Charles, and Shimazaki Ryusuke, *Weatherman*, performance at SuperDeluxe, Tokyo, 2011. Courtesy of the artists.

collaborative work, *Weatherman* emerged through on-site discussions regarding the effects of the events, their representation in the media, and how these impacts were felt on a personal level.

It was not just collaboration between people; it was also a coproduction of place. Place is as much a conceptual as a geographic entity for artists: "The artist does not simply dwell in a place but collaborates with place" (Papastergiadis 2011, 88). Through the use of digital technologies and social networking systems, different forms of community and participation emerge that are not wedded to specific sites or territories. Rather, they reimagine relations to place that include both physical and online place. How we navigate the world and our experience of it increasingly occurs, at least in part, through digital media (Pink and Hjorth 2012). It is therefore *part of*, not separate from, regional place.

Conclusion: From Critique *of* to Collaborative Engagement *With*

In chapter 6, we explored the relationship between art and the environment, showing how artists in the region have developed critical responses to environmental issues. In doing so, we to some extent revised the ways that nature is portrayed in dominant discourses. In this chapter, we have shown how other artists are moving critique further by exploring models for collaborative cross-cultural art projects in the Asia-Pacific. Such projects both build new relationships through collaborative practices and use networked and social media to reconfigure art practice and audience engagement. This opens up a position from which we can also consider the role of mobile media in these relations of place, particularly how, as a vernacular media practice, online mobile culture is changing what "public" and "intimacy" mean and the entwined relations between the personal and the political. In the next chapter, we discuss examples of arts practice that explore new models for public and private engagement and how they might be conceived as generating new forms of intimate publics.

8   Greening Intimate Publics

In this chapter, we discuss how mobile media might be engaged to generate ways of resolving the disconnections that, we have argued, frame much of the existing methodological paralysis around the question of how to address questions of climate change through art. In the preceding chapters, we have shown how there are two key disconnects in the ways that mobile media and arts practice are situated in the Asia-Pacific. Here we discuss the potential of mobile media art to question waste and the impact of the digital on the material.

The first key disconnect is the failure of existing literatures and environmental discourses to make any sustained and substantial critique of the deeply worrying and dangerous ironies around the region's status as a core producer and consumer of mobile media technologies. The dark side of the apparently global need to consume, discard, and replace predictably cyclically obsolete digital technologies has deeply penetrated the region. This has been witnessed in public scandals relating to the working conditions experienced in the production of globally leading digital technologies in the region in recent years (see, for example, Qiu 2011).

Yet, at the same time, the region is also home to the world's first example of mainstream mobile Internet, which emerged in Japan more than fifteen years ago, as well as strong cultures and economies of digital innovation and civic participation. These different manifestations of the politics, economies, and social and cultural worlds that mobile media technologies and apps sustain in the Asia-Pacific are mutually interdependent and, in being such, part of a destructive cycle that is ultimately damaging the environment. Nevertheless, discussions about the role of mobile media and their relationship to the environment have remained relatively cursory. While, as we have shown in earlier chapters, there has been no shortage of digitally mediated activism or civic engagement across the region with respect to other issues, there is little public critique or participation in debates about the role of digital media in the

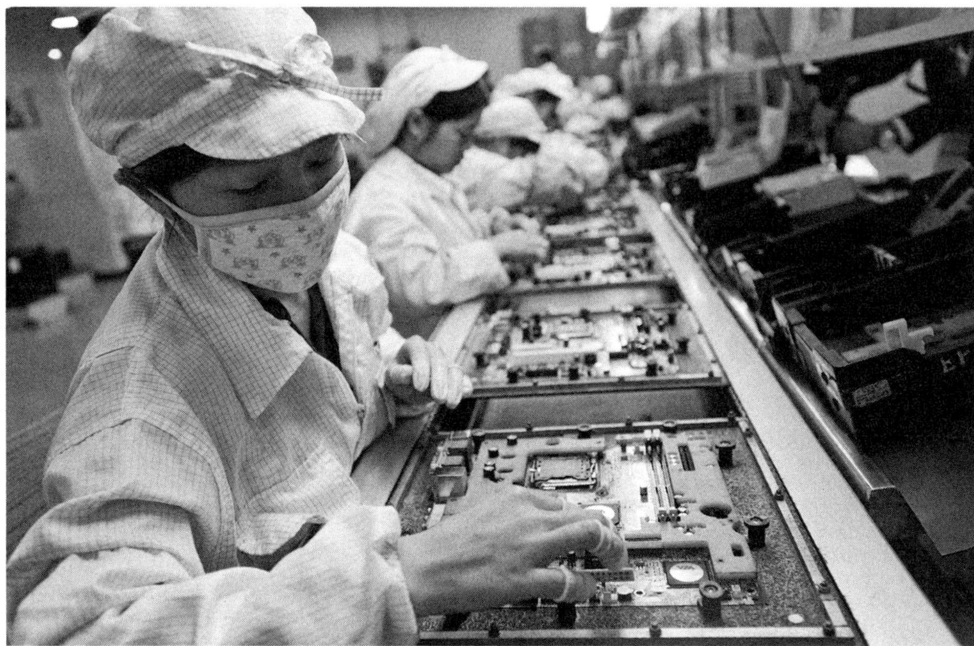

Figure 8.1
iPhone manufacturing in Shenzen, China. Photo: Getty Images.

ironic relationship between economic growth and environmental degradation in the Asia-Pacific.

The second disconnect occurs between the need to acknowledge the capacity of arts practice, and of public art in particular, to generate new forms of public participation and debate about environmental issues, and the extent to which this capacity is being formally acknowledged in key and influential sectors of the art world. Connections between art, media, and the environment can provide new ways of reimagining the environment and also act as a pathway for revising the increasingly blurred relationship between new media and art. The discomfort between art and new media practice was demonstrated by Claire Bishop (2012a) in her discussion of the two fields. There, Bishop did not account for the ways in which new media had become an embedded part of many visual artists' conceptual premises, that is, how new media are not just a technology to be used by artists but also a way of framing cultural practice.

What becomes apparent is that the distance between the two fields is no longer tenable in practice. We thus need more complex conceptualizations of the role of art in contemporary online and mobile media. In previous chapters, we cited the work of artists from the globally celebrated Ai Weiwei and Cao Fei to groups such as Command N and the Candy Factory. The region has been home for decades to synergies between art, media, and place and so provides an ideal backdrop to move toward the engagement of mobile media's connections to place.

While mobile media and smartphones are becoming increasingly ubiquitous in everyday life in the Asia-Pacific, they are dislocated in public and academic discourses and also from much of current arts practice dealing with environmental issues. In this chapter, we build on approaches to mobile media developed in cultural studies, engaging the concepts of the intimate, public, personal, and political (Hjorth and Arnold 2013) to propose new connections within these fields. We examine two themes. First, we look at how the potential of mobile media to render the intimate *public* and the public *intimate* is shaping, and being shaped by, place, art practice, and politics. Second, we address how the new models for engagement, distribution, and participation that are promised by mobile technologies and online platforms are changing how art is practiced and how audiences participate.

The ubiquity of social media within contemporary everyday life has created new opportunities to bridge the traditional divisions between art and new media. However, the extent to which they have been addressed is uneven. For example, Bishop (2012a) sees digital technology—whether through its embrace or its denial—as being central to contemporary debates. Her own interpretation proposes the use of digital media as a technology for artists to use, rather than as a mode of critical practice. However, it is not simply the presence of digital technologies and apps that it is difficult for artists to avoid. Rather, we argue that the potential of media for generating forms and feelings of intimacy in public settings and vice versa should be seen as an invitation for artists to consider the possibilities this offers for public engagement with and in arts practice, events, and platforms. In earlier chapters, we conceptualized how what we call "intimate publics" (Hjorth, King, and Kataoka 2014), which can emerge as both public and intimate domains, are reconfigured in a digital-material environment. Here the online and offline are part of the same world.

We have also argued that the ways in which people move through such worlds can be conceptualized as a form of digital wayfaring (Hjorth and Pink 2014). This concept enables us to think about how mobile media and smartphone users repeatedly traverse environments that are simultaneously online and offline, and in doing so traverse new configurations of public and private domains. These configurations do not so much challenge more traditional forms of arts practice as provide an opportunity by offering ways to engage publics. They do so by creating meeting points with everyday smartphone users as participants in public art projects as they navigate digital-material worlds where the public and intimate are entangled.

These forms of public–intimate entanglement are already evident in arts practice in the Asia-Pacific. For instance, the work of collectives like the Australian Artist as Family is indicative of the significance of the ways in which the personal and intimate become implicated with the political and social through arts practice (http://theartistasfamily. blogspot.com.au). As this collective puts it, "Artist as Family practice a unique form of performance art. Performances comprise how we live, get our food and move around; performing low-carbon modes of life making" (2014, n.p.). From edible work (*Food Forest* [2010], exhibited at MCA in Sydney) to collecting anthropogenic waste along the coastline and throughout the city of Newcastle (*17 Days* at the This Is Not Art Festival), Artist as Family is committed to embedding social practices of sustainability within art

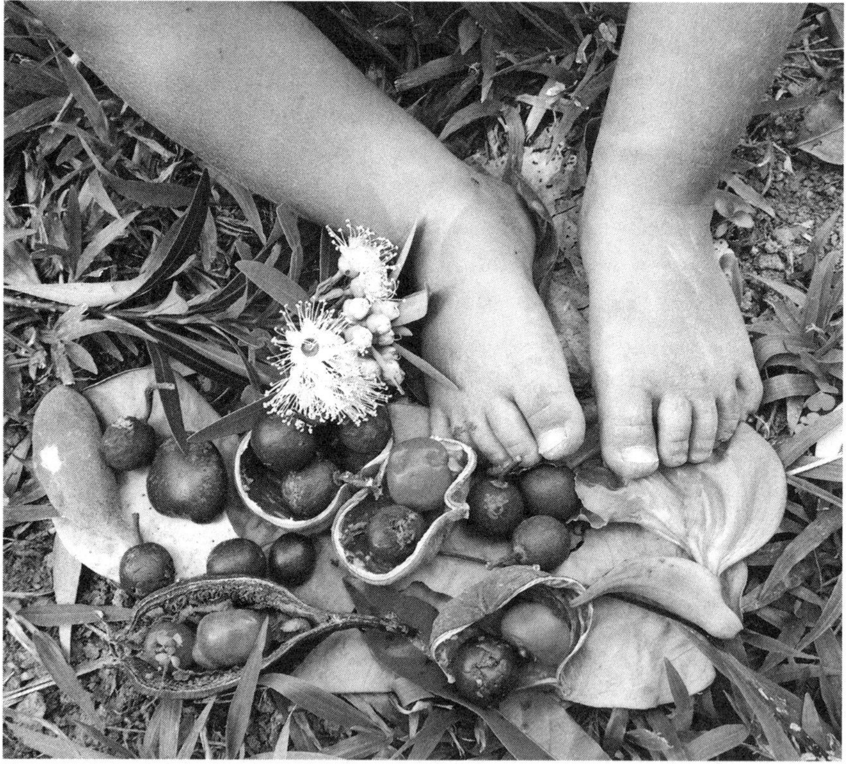

Figure 8.2
Artist as Family, *The Art of Foraging for Food* (ongoing). Photo: Artist as Family.

contexts. Through cultivating a sense of intimate publics, Artist as Family explores the relationship between biodiversity, sustainability, art, and the environment. For the artists of the collective (who are an actual family of two parents and two children, including Patrick Jones, Meg Ulman, and Zephyr Ogden Jones), the role of art is to intervene in the everyday and make an impact socially and politically.

Within the art world, as we have shown already, the Asia-Pacific is a leader in the uptake of social media to facilitate local and global connections and engagement between artists, publics, and institutions. Across the region, these media forms are being used in an explicitly political manner that challenges the nonpoliticized discourses of participatory and collaborative practice that dominate Eurocentric approaches. It was not by accident that one of the leaders in locative art, Rafael Lozano-Hemmer, conducted seminal works in the region. In particular, his *Relational Architecture 1*, which featured at the opening of the YCAM Center in Japan (2003), required that participants' phones have cutting-edge locative media capacities. Only in Japan, with its innovative mobile media, could such an event have taken place. More generally, however, artists have been slow to adopt such technologies innovatively. They continue to employ online media to facilitate the production and distribution of information rather than to encourage user participation.

For instance, Japanese new media groups like Candy Factory or the Korean American collaborative team Young-Hae Chang Heavy Industries have been seen as indicative of the first generation of Internet artists (Greene 2004). But a closer look at their work shows that they use online environments and mobile technologies as an alternative to the preferred offline space. The aesthetics of both groups epitomize the flash style of Web 1.0; and yet in an interview, members of the Candy Factory describe the online as a default space that runs parallel to their offline events, suggesting that the group is using the online as a site for dissemination rather than as part of the conceptual development of their art (Hjorth 2013). In contrast, regional precursors such as Japan's Dumb Type in the 1990s and Korea's Nam June Paik in the 1970s were quick to embrace the popular technologies of the moment as a commentary on broader sociopolitical trends.

Other early interventions also began to move between online and offline spaces. For instance, the Korean collaborative group INP began its foray into online–offline spaces with Blast Theory–type performances (using mobile technologies to explore hybrid realities) and then quickly moved to software and hardware mobile hacking workshops to unpack the political dimensions of media practice. INP saw mobile technologies as symbolic of South Korean politics and media ecology. By deconstructing these symbols, the group provided a space in which participants could question and

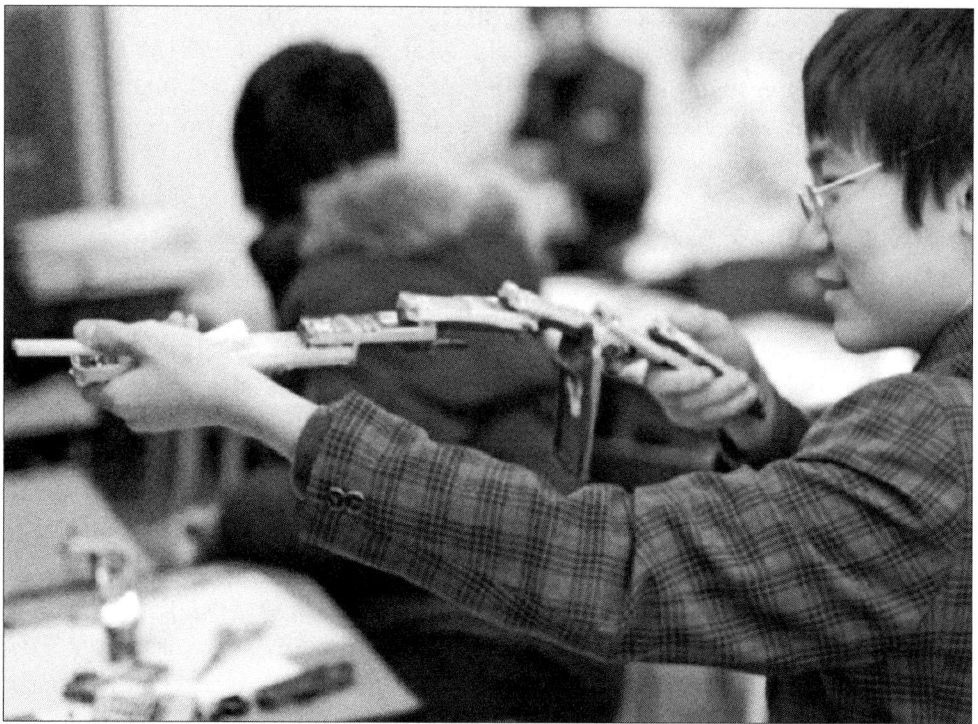

Figure 8.3
INP mobile hacking workshop, Seoul, South Korea, 2005.

challenge hegemonic technnonationalism (Hjorth 2008a, 2010). This questioning of technnonationalism was continued through a series of INP's mobile hacking workshops.

As one of the key regions producing mobile technologies for global consumption, the Asia-Pacific's politics of media practice, software, and hardware provide a rich and complex context for conceptual and aesthetic exploration. Command N's projects in Japan offer a fascinating example of a trajectory that has brought the public and private together using media technologies. For instance, Command N's *Akihabara TV* project in 2000–2001 placed video works on public and private screens in Tokyo's Akihabara (Electric City). By deploying public screens for these subversive artworks, the group demonstrated how art can make intimate yet public interventions within the everyday.

More recently, Command N has sought to use mobile and online media via the WAWA Project to engage audiences in a synergy between art, media, and the environment through community-based practice. The 3/11 earthquake, tsunami, and nuclear disaster in Tokyo (discussed in chap. 4) had an enormous impact on the types of projects WAWA focused on. As the horrific events unfolded, mobile devices were crucial in material and immaterial ways. In the interviews that Hjorth and Kim (2011) conducted in Tokyo after 3/11 about the role media played in crisis management, many people spoke about the ways in which, as the technology failed functionally, it grew in symbolic importance. One respondent spoke about cradling her *keitai* (mobile phone) as she went home, as if it contained all her friends and family. For others, when the technology did work, it made them feel bombarded and overwhelmed as they attempted to grasp the reality of the situation while in a fog of shock and frozen by grief.

Social media are increasingly used for political activism (Christensen 2011; Diamond and Plattner 2012), not least in the Asia-Pacific (Postill 2014). The ongoing 3/11 protests in Japan relied significantly on social media to publicize and represent issues relating to the nation's reliance on nuclear power. The mainstream press minimized the reporting of the 3/11 event and its attendant public protests, even though it is arguably one of the more significant forms of public protest in Japan since the 1970 campaign against the Japan-U.S. Security Treaty (Slater, Nishimura, and Kindstrand 2012).

Significantly, *No Nukes Girl* (1997), an artwork by Yoshitomo Nara (unrelated to the events of 3/11), was co-opted as the pinup image for the protests against nuclear power, and when worn and used by demonstrators as sandwich boards, it also became an embodied relation of media, art, and environment, as the image was quickly circulated for download through social media, with the artist giving his permission for its use via Twitter. Local government launched public appeals for aid on YouTube, and individuals used blogs and other means to disseminate alternative information regarding radiation levels in urban areas (Slater, Nishimura, and Kindstrand 2012). Significantly, these uses of digital platforms combined with traditional media forms and outlets to share sources and materials. In the next section, we focus in more detail on the theme of mobile intimacy to suggest how it has already been engaged, and to explore its potential in generating new forms of critical public engagement with environmental and climate change issues through art.

Creating Mobile Intimacy through Art

As we discussed in chapter 2, a theory of movement is integral to understanding not only mobile media but also more broadly how we as humans live in the world and the processes through which things and intangible flows circulate, are distributed, and intersect. Mobile art practice, which uses movement as a way of engaging with publics, can also be used to engender a sense of closeness or intimacy both to the artifact itself and to the issues it raises. For example, the Malaysian-born, New York–based artist Tattfoo Tan's socially engaged artworks aimed to create new forms of environmental awareness by mapping them on locally significant and emotive issues. His *S.O.S.* (Sustainable. Organic. Stewardship) *Mobile Garden* used found shopping carts and baby carriages and refitted them so that they could become "mobile gardens." Tan would then push the mobile gardens around and leave them in strategic points throughout New York. Tan has described the project thus:

Mobile Garden is also inspired by the ghost bike memorial that is around the city to mark the place where a bicyclist was killed. Now Mobile Garden can be a marker, a reminder to the scarce of land, its right and usage. Maybe we can think about how to repurpose our urban environment. … An edible garden is also a sign of our time. The lack of food in the world. The tainted food supply, caused by industrial gardening. The rise of the organic movement. The cost of producing and transporting (carbon footprint) the crops and produce, and the effect it had on the environment. The disease of overconsumption and the need to purchase inferior quality product that will be thrown away is indeed needing to be eradicated. (Brown 2014, n.p.)

Tan's *Mobile Garden* provides us with one way to understand the tensions around sustainability and being "mobile" in a contemporary context. By moving and camping in public sites, Tan's gardens become intimate entanglements within people's everyday lives. Tan chooses strategic placements for his mobile gardens, such as areas in dense urban areas where having a garden is not an option.

However, the idea of being mobile has also been interpreted in terms of what have been called nongeographic mobilities (Hannam, Sheller, and Urry 2006). For instance, the notion of "being moved" has long been attached to emotion (Lasén 2004), particularly labile or volatile emotion, but also to its more sanguine forms. Indeed, by bringing together the emotional, geographic, and nongeographic dimensions of mobility, we can begin to understand how the "mobilities turn" (Hannam, Sheller, and Urry 2006) and the "intimate turn" (Goggin 2011), when considered together, offer a framework for considering the potential of mobile media for generating intimate publics through arts practice. As Anthony Elliott and John Urry (2010) note, multiple forms of mobility are transforming our everyday lives. From travel, transport, and digital technologies to global consumerism, mobility has been a key feature of twenty-first-century "postcarbon" lifestyles. When we consider the various forms of large-scale global mobility that characterize our era—mobile people, mobile ideas, mobile resources, mobile markets, mobile labor, and mobile capital—digital technologies and the Internet are clearly implicated (Castells 1996).

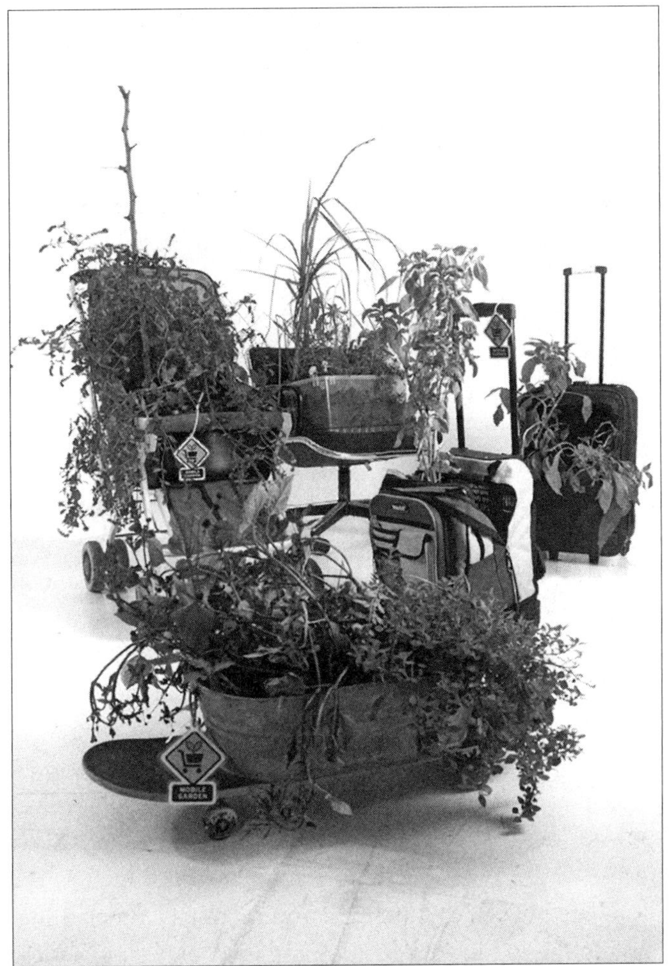

Figure 8.4
Tattfoo Tan, *S.O.S. Mobile Garden*, 2009. Courtesy of the artist.

It follows from this argument that mobile practices, mobile media, and mobile publics should be the vehicles through which the environmental and climate change issues raised by this media-hungry context (Maxwell and Miller 2012) might be addressed. The work of artists like Tan shows how art can reflect on debates about the impact of mobility (especially enforced) and sustainability within contemporary life. Later we consider in more detail how such issues have been addressed through media art. First, however, we further situate the role of mobile media in such processes by outlining how the forms of mobile intimacy they generate in everyday life also offer routes through which to consider their potential as technologies for public art interventions.

Mobile and smartphone technologies, apps, and the social media practices they are associated with are also implicated in forms of smaller-scale mobility. This includes the daily mobility of family members, the mobility of a given workforce, and the mobility of our personal arrangements in time and place (Hjorth and Lim 2012). As technologies of propinquity (temporal and spatial proximity), they are both instrumental in, and symbolic of, new erosions of the boundary between family and profession, public and private, work and leisure (Wajcman, Bittman, and Brown 2009). This is partly connected with the widespread use of locative media within the everyday mobile media experience, which has shifted the manner in which we imagine and navigate the online in conjunction with physical spaces. One way to conceptualize this straddling of digital and material experiences of copresence is to focus on the notion of mobile intimacy. Personal mobile media present us with boundaries between online and offline worlds, public and private spaces, an inner life and a social performance, and at the same time enable us to skip back and forth across those boundaries at will. At the same time, the forms of copresence they engender create emotional and social experiences. This overlaying of the material-geographic and the electronic-social sets the conditions for what we call *mobile intimacy*.

There is a rich historical context for understanding how the relationship between mobility and intimacy has changed across contexts, platforms, and technologies over time. As Timo Kopomaa (2000) observes, today's personalized mobile media can be seen as an extension of nineteenth- and twentieth-century mobile media such as the pocket watch and the wristwatch, which both personalized and mobilized time telling. Similarly, technologies such as mobile media reconstruct earlier copresent practices and interstitials of intimacy: for example, the metaphor of "posting" in SMS (short message service) both reflects and subverts earlier epistolary traditions such as nineteenth-century letter writing (Hjorth 2005). Likewise, scholars such as Christian Licoppe (2004) have argued that new forms of copresence, such as e-mail, are linked to earlier practices of intimacy through the absent presence of media such as visiting cards.

However, some striking differences emerge. Mobile media can be considered part of a shift in conceptualizing and practicing intimacy as no longer a "private" activity but a pivotal component of performativity in the public sphere. Intimacy has taken on new geo-imaginaries, most notably as a kind of "publicness" that is epitomized by the mobile phone (Fortunati 2002). The role of the mobile phone as repository for the personal and pedestrian has been foregrounded not only in Japan (Ito, Okabe, and Matsuda 2005) but also in the region more generally (Hjorth 2009). In the next section, we advance this discussion further by considering how intimate and social publics have been constituted through arts practice, and the implications of these examples for the role of mobile media in arts practice.

Intimate and Social Publics

The notion of an intimate public might be interpreted as being manifested through Asia-Pacific artworks in a number of ways. One example is the work of the Beijing-based Liu Ding's phone music pieces. The work *1999* (2014) comments on 1990s popular culture in China. It is an installation that consists of recorded quotes and pop music of the 1990s, which can be listened to on the telephone. In *1999* (2014), audiences can pick up a landline phone (pre-mobile phone) to listen to music and quotes. A hybrid between headphones and mobile phone, the telephones are instead stationary receptors that transport the listener to another time. We are reminded of the important

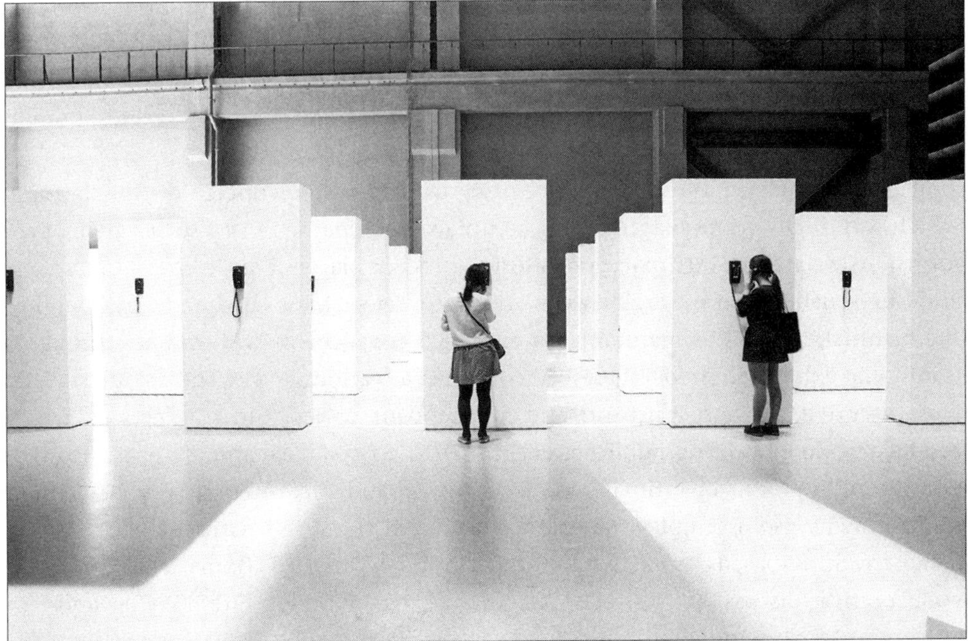

Figure 8.5
Liu Ding, *1999*, 2014. Courtesy of the artist.

work by Rey Chow (1997) in her study of the Sony Walkman in China, which explored making personal and private the act of listening, working against the history of listening always having been public. Liu Ding's *1999* inverts this history of listening by deploying a nonmobile media (the landline) to blur the public with the private, the intimate with the social, in ways that are both culturally specific and also cross-cultural. Rather than listening being mobile and private as with the Walkman, audiences are frozen in space and also in time (stuck in the nineties). Playing with old and new media also evokes the importance of copresence (especially across time) while being both mobile and still.

If, as we can interpret from this example, arts practice can deploy media to generate intimate publics through media uses that are at once personal, mobile, and social, then we can also see how such work situates intimacy in a fluid interpersonal and conceptual space in which binaries and boundaries are broken down and refigured. As social, locative, and mobile media become more pervasive, different modes of using these media mean that publics are increasingly constituted through the generation of *intimate*, rather than networked, relationships and ways of connecting people. As we have argued, the notion of mutually exclusive private spaces and public spaces is problematized by sociotechnologies that enable public interaction from a position of privacy and by switching between the two. In particular, the concept of intimate publics helps us begin to understand how the entanglement of art and media practice can lead to productive ways to rethink public engagement with art and the critical issues it might seek to address.

These uses of media in arts practice bring us back to the concept of the platform, which was discussed in chapter 5 (for biennials) and chapter 4 (for social media platforms). As we argued, biennials generate a significant place for art in the public sphere and can be defining of the region as an arts space. Yet we have much to learn from how the biennial, which has maintained a distanced engagement with publics through a traditional exhibition model, has been contested at various levels. These contestations have occurred through artist-run and independent projects and collectives, which produce works that bring publics much closer to arts practice and are geared toward generating the forms of intimate connection with audiences that are associated with mobile media use. Examples that have emerged in the Asia-Pacific include Platform Seoul, the Seoul-based annual exhibition by Sunjung Kim; and Utopia, a cross-cultural, collaborative discussion between Asialink, Tokyo Wonder Site (TWS), and the Singapore Art Museum.

The relationship between art, publics, and intimacy is not new. For Mami Kataoka (2014), the idea of "intimacy and distance with the public" can be found in the practices of Ai Weiwei, Lee Mingwei, and Allan Kaprow. Drawing on the work of 1960s artists, Kataoka shows how art has long been haunted by the role of media and its relationship in constructing art publics and intimacy. This haunting takes on particular art and intimate public specters in post-3/11 Japan. Kataoka's argument about the importance of connecting contemporary debates about art, media, and the environment in Japan with their history in social movements is also pertinent. In the next section, we discuss the WAWA Project, which is informed by both a Japanese-specific and a global realignment about the importance of art in connecting the community with the environment. To do so, we begin with a brief history of environmentalism in Japan.

WAWA: A Contextual Example

According to Japan's Ministry of Foreign Affairs, environmentalism in the nation began during the Meiji period (a period marked by great industrialism) and went through two key stages after 1960s urbanization. The first Japanese environmental movement emerged in the 1970s, exemplified by the Minamata and the Love Canal incidents. Together with rapid motorization, air pollution via photochemical smog became one of the major environmental issues. People soon came to recognize that the wrongdoers were not only companies; consumers also played a role in the problem. The second stage of environmentalism, represented by antidevelopment and ecological movements in the 1980s and 1990s, focused on identifying ecologically sound lifestyles.

In the 1990s, pushing back against the rapid industrialization and urbanization of the 1970s and 1980s, local grassroots and community-based environmental movements, dubbed the Jyumin-Undo (local residents' campaign), emerged (Mitsuda 1996). Writing a few years after the landmark United Nations Conference on Environment and Development (UNCED) of 1992, Jonathan Taylor argues that, since the early 1990s, "Japan has attempted to position itself rhetorically as a global environmental leader"; however, for Taylor, this position is underscored by the reality of the "Japanese model of development" as "linked to Asia's continuing environmental crises" (1999, 535).

And yet the traditional notion of *mottainai* ("what a waste"), while repressed until the 1980s economic bubble burst, has continued to play a key role in everyday Japanese life (Masters 2008). The significance of *mottainai* is also evident in the work of many contemporary Japanese artists, such as Akiko and Masako Takada, whom we discuss later in the chapter. In 2000 the Basic Law for Establishing the Recycling-Based

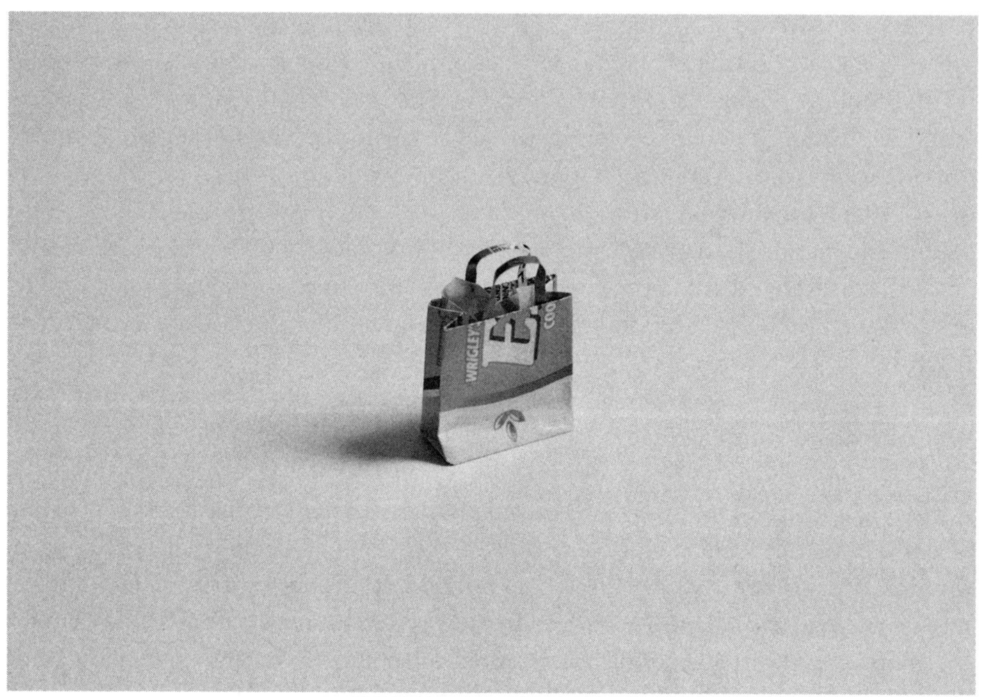

Figure 8.6
Akiko and Masako Takada, *Shopping Bag*, 2008. In its transformation of everyday materials, the Takada twins' work exemplifies the Japanese principle of *mottainai*. Courtesy of the artists.

Society came into effect. Given Japan's tendencies of mass production, mass consumption, and mass disposal, the law aims to shift emphasis toward a reduction of environmental loading.

While grassroots, community-driven environmental groups have grown over the past three decades, it was in the wake of 3/11 that new forms of environmentalism began to emerge that involve forms of protest entangled with the digital. As David H. Slater observes in his analysis of Japan post-3/11:

The threat of nuclear radiation and critiques of the nuclear industry have been skillfully politicized in ways that have led to the largest set of demonstrations in Japan (with the exception of Okinawa) since the US-Japan security treaty protests of the 1960s and 1970s. These protests have been based in Tokyo, utilizing urban networks of activists who have provided the digital framework for organization that has brought together an older generation of anti-nuclear activists, young families, hip urbanites, office workers and union protesters. (2011, n.p.)

And yet, in the case of 3/11, women (and especially mothers) have become the driving voice for antinuclear protests in Japanese society and policy. As Slater observes, "Women, and in particular, mothers, have been quite active in radiation measurement, calls for contaminated soil removal, and efforts to secure safe food since the early months of the crisis. Today, perhaps more than any other group, they have emerged as particularly effective anti-nuke spokespersons." This movement has been dubbed the "Women from Fukushima against Nukes" (Genptasu iranai Fukushima kara no onnatachi 2011). In another article, written jointly with Nishimura Keiko and Love Kindstrand, Slater highlights the pivotal role that social media has played in new forms of environmental activism post-3/11.

The fervor, organization and momentum that culminated in the events that occurred up to and around the September 19, 2011, march of 60,000 people in Tokyo, the largest such gathering since the 1970s AMPO campaign against the Japan-U.S. Security Treaty, were seen by many of us as examples of the unexpected power of social media to provide alternative narratives that could give rise to significant political action. (Slater, Nishimura, and Kindstrand 2012, n.p.)

Indeed, the ambient, intimate, and copresent characteristics of social media have afforded young and old alike with new ways to express and engage with politics. Social media such as Twitter, Instagram, and Line are viewed as both political and personal (Hjorth and Arnold 2013). It is this engagement with politics, especially with green issues, that the next section explores by taking a closer look at collectives such as WAWA and 3331 Arts Chiyoda, as well as contemporary artists in Japan.

For the Japanese artist-run collective Command N, art and media have always been productively entangled. Established in April 1998, Command N conceived its name after Macintosh's keyboard shortcut. Headed by the artist Masato Nakamura, Command N is famous for its highly successful *Akihabara TV* project, which featured works from artists like Patricia Piccinini being projected onto hundreds of screens inside and outside stores in 2000 and 2001. When 3/11 occurred, Command N developed the WAWA project to connect artists, activists, and the general public to collaborate on creative ways to reinvent a sense of community and belonging. WAWA is indicative of the types of collaborative and participatory art activities emerging in the region, which engage complex relations between material and immaterial place formed through intimate publics.

WAWA means "I am," as well as circle, unity, and togetherness. Dubbed a "social creative platform," WAWA created a network, a series of workshops, and exhibitions that brought together various figures who have inherited the legacy of living, breathing cultures from the quake-stricken Tohoku area, including sake-brewing artisans, Shinto priests, salt-making studio craftsmen, and nonprofit organizations that implement art projects in the rural provinces.

Through a series of activities between artists and non-artists, WAWA highlights the importance of the social in locating art practice. Far from the "aesthetics of the social" that Claire Bishop (2004) identified as problematic within relational aesthetics, projects like WAWA show how art plays a pivotal role in instigating change, especially in times of crisis and trauma. As Kataoka notes, many felt compelled to make art that engaged with the realities of post-3/11 as well as helping to raise awareness and funds for the victims. She argues that this tension around intimacy and distance is not just a new phenomenon today. Rather, through various practices from Kaprow onward, she argues:

If Tiravanija's projects, which involve offering meals to guests at opening parties and receptions, inherit the legacy of Kaprow's Happenings, Lee's one-to-one artworks might be interpreted as having an affinity with Kaprow's Activities. The discrepancy that exists between these types of relationships and the desire for a one-to-one relation are fundamentally connected in terms of the nature of the essential desire that underwrites them both, even in a contemporary context characterized by advanced media technologies. (Kataoka 2014, 129)

For WAWA, platform as a discursive device is used to reconnect activities, art, and the community through a variety of modes. According to their website:

[WAWA] engages closely with activists in affected communities, maintaining a base for their activities, while transmitting information about their endeavors through online and offline publications. The name of this project came from our commitment to focus on and visualize the individual person who is working in an effort for recovery. By saying WAWA, this translates as "I am," conveying our wish to activate one person at a time. Collectively individuals with much competence become "we," generating greater power to move forward. As the name, Social Creative Platform for Opportunity, states, we want to multiply opportunities that boost the sustainable human energy to rebuild these affected areas. (WAWA Project 2014, n.p.)

What WAWA identifies so vividly is that the relationship between media, art, and the environment is an urgent one. Art cannot sit outside the trauma of 3/11; instead it is a way for bringing together both mobile and immobile intimate publics. Through the entanglements between art, media, and the environment, we can mobilize intimate publics to share, collaborate, and create in ways that both reflect previous art strategies and also provide new ways to act for change.

As a collective committed to socially and politically engaged projects that explore the relationship between community, place, and the environment, WAWA has numerous ongoing projects. One project is called Musubu, which is about "connecting people, communities, and the world" (2014, n.p.). Many of the collective's projects are dedicated to the idea of *machizukuri*, which means "community building" (*machi* is a noun meaning "community" or "neighborhood," and *tsukuru* is a verb meaning "to build"). Another incubator is MMIX Lab, which seeks to bolster community creativity through art and multimedia.

One project of the lab is *3.11 Memorial Project*, which seeks to "preserve the surreal sights of ships landed in unimaginable places and public signage twisted and distorted by the tsunami." In the same theme, *Sakura Project* "intends to visualize the reach of the tsunami through a line of cherry blossom trees, so its destructive force may not be forgotten. These projects take the force of nature and transform it into the creativity of regeneration" (WAWA Project 2014, n.p.). Many of the projects seek to learn from the trauma and to develop stronger community-based engagement, as well as creating community-built monuments to memorialize the disaster.

In addition to WAWA, which seeks to transform the trauma of 3/11 into a community-building exercise through art and media collaboration, founder Nakamura also directs 3331 Arts Chiyoda, a space committed to socially and community engaged art projects (http://www.3331.jp/en). The space focuses on inner-city revitalization and developing new scenarios for engaging with the complexity of the urban situation. Based in a school and having access to gallery spaces and a residency program, 3331 is one of the few existing artist-run spaces. While collectives like WAWA and 3331 Arts Chiyoda have explored more overt community-building (*machizukuri*) projects, many contemporary Japanese artists are investigating the relationship between the environment, technology, and representation through deploying the *mottainai* (not wasting) by reinventing everyday materials in new ways.

The deployment of tensions between the intimate and the public, and its manifestation within representations of the environment, can be found in numerous Japanese artists' practices. The aforementioned artist Takahiro Iwasaki (see chap. 4) is indicative of a different type of engagement with the intimate and mundane as a means to reflect on urban interventions and the environment. Within the same generation of Japanese artists, we can find similar artists probing the intimate and social in new ways to reinvent how we reflect on the environment. Okinawan-born, New York–based Yuken Teruya uses traditional craft skills to critique consumer culture. In *Dessert* (2009), Yuken deploys sweet materials like icing, gum, jelly, sponge cake, and cream to create a diorama that reflects the paradigm shift from the geocentric model to the heliocentric model—in other words, from envisioning the earth as flat to understanding the earth as round. In *Minding My Own Business* (2011), Yuken used newspaper to make a miniature origami-like forest to comment on the 3/11 disaster.

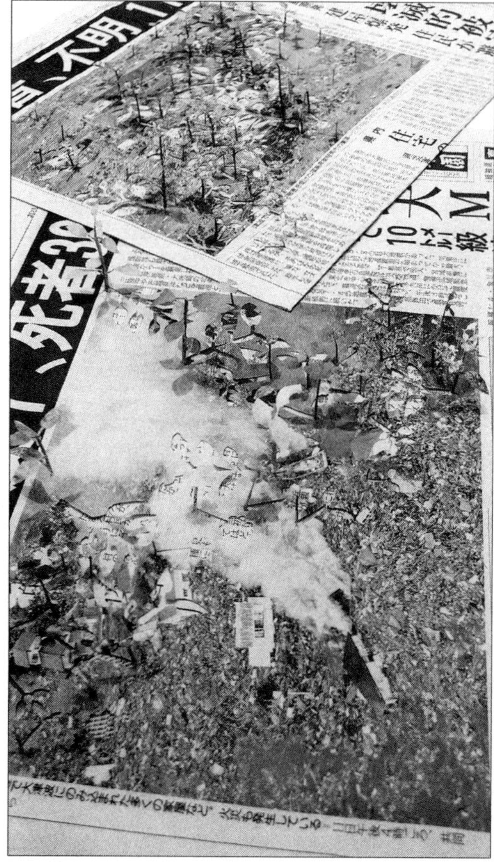

Figure 8.7
Yuken Teruya, *Minding My Own Business*, 2012. Courtesy of Yuken Teruya Studio. Photo: Yuken Teruya Studio.

The Japanese collaborative twin artists Akiko and Masako Takada also deploy intimate registers to explore notions of the environment as a lived experience and representation. *Airline Route* (2009) is a globe that shows the cartographies and lines of flight paths. The twins often use materials such as sugar and paper to reinvent how we imagine and visualize the environment in both its natural and built forms. By combining craft skills with consumer culture, much like Yuken, the Takadas provide playful and engaging works that allow us to rethink conventions and normalizations around what is natural and what is constructed. The Takadas take common objects and reinvent their use—much like Tom Friedman (who takes everyday objects and reinvents them

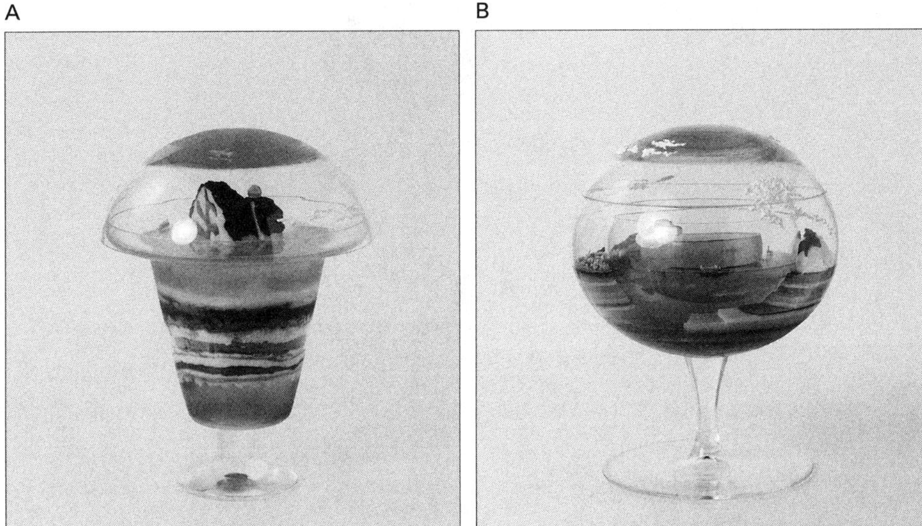

Figure 8.8
Yuken Teruya, *Dessert*, 1999–2006. Courtesy of Yuken Teruya Studio. Photo: Yuken Teruya Studio.

by playing with the size and scale). However, there is something uniquely Japanese in their deployment of scale and its relationship to movement and stillness in the environment.

The Japanese artist Yoshihiro Suda further pushes the entanglement of intimacy, nature, technology/craft, and the environment. Suda meticulously carves wood into flowers and weeds, infusing the natural with the handmade. He paints the wood to further return it to its "natural" state. His small sculptures are never placed on pedestals; rather, they are arranged on the floor as one might find flora in nature. Here, toying with nature is extended to reflect on the politics of representing nature. Suda's constructed nature is popping up in supposedly highly constructed sites, demonstrating the power of nature to intervene and disrupt technology within the everyday.

Conclusion: Mobile Publics for Intimate Engagement

In this chapter, we have discussed the ways in which the social and the intimate, the public and the political, have become and continue to be entangled. Through the rubrics of mobile intimacy and intimate/social publics, we have sought to revise how we might interpret contemporary intersections of art, media, and the environment in

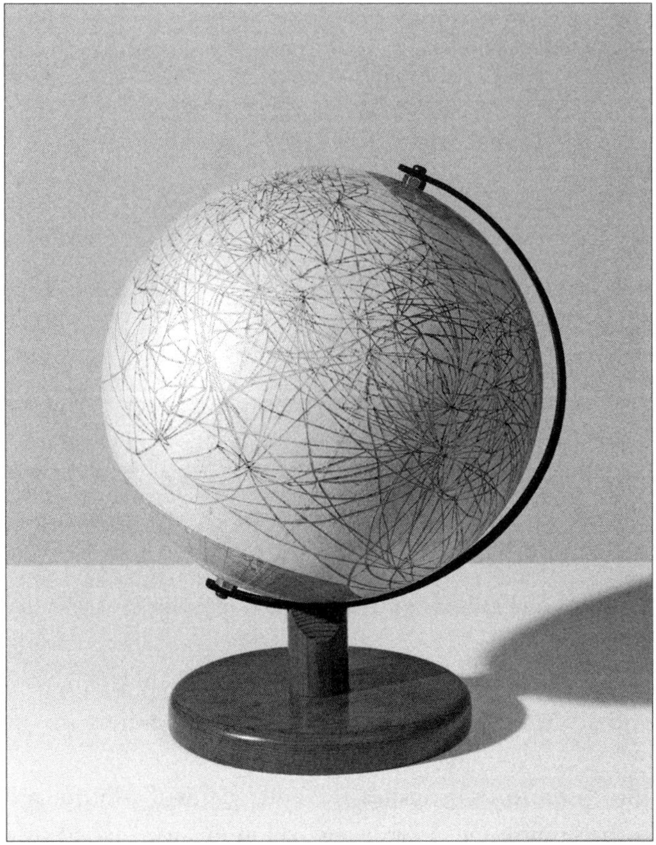

Figure 8.9
Akiko and Masako Takada, *Airline Route*, 2009. Courtesy of the artists. Photo: Hideto Nagatsuka.

the region. Given the impact that social and mobile media have had in political activism, as well as a prevalence of debates about the dumbing down of the social into a prefix for social media and its attendant banalities, we have reflected on how these shifts in the role of media, and its mediation of intimacy, politics, and notions of the public, are reshaping art and its place in the environment. Through exploring artist collectives and artists, especially in the context of a post-3/11 Japan, we have reflected on some of the ways that art, media, and the environment can be entangled to revitalize and engage community participation.

Since the mobilities scholar Mimi Sheller first defined the notion of the "mobile public" (2004, 39), mobility and publics have taken on more complexity. Sheller

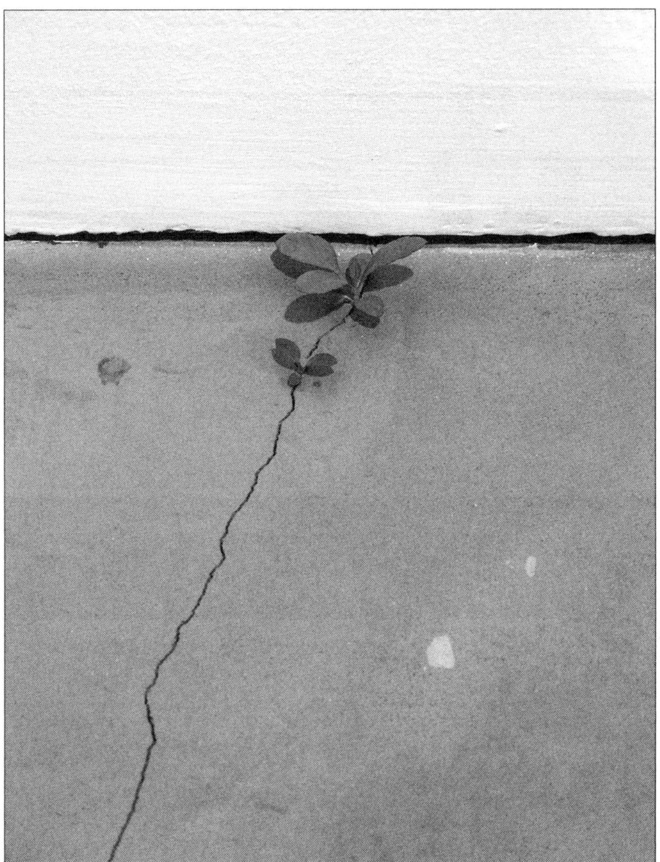

Figure 8.10
Yoshihiro Suda, *Weeds*, 2008. Paint on wood. © 2008 by Yoshihiro Suda. Courtesy of Gallery Koyanagi Tokyo.

discusses the mobile transformations of public and private life as a way to conceptualize beyond the network perspective (as proposed by the likes of Castells). Like us, contesting the ways in which network metaphors have come to dominate discussions of digital media, Anne Galloway has suggested reworking the notion of "mobile publics" in a way that also helps us to understand contemporary art and politics in an age of social and mobile media. Galloway has suggested that "given the imperatives to locate and connect that are embedded in otherwise diverse technologies, it should come as no surprise that today's wireless and wearable devices and applications offer unique

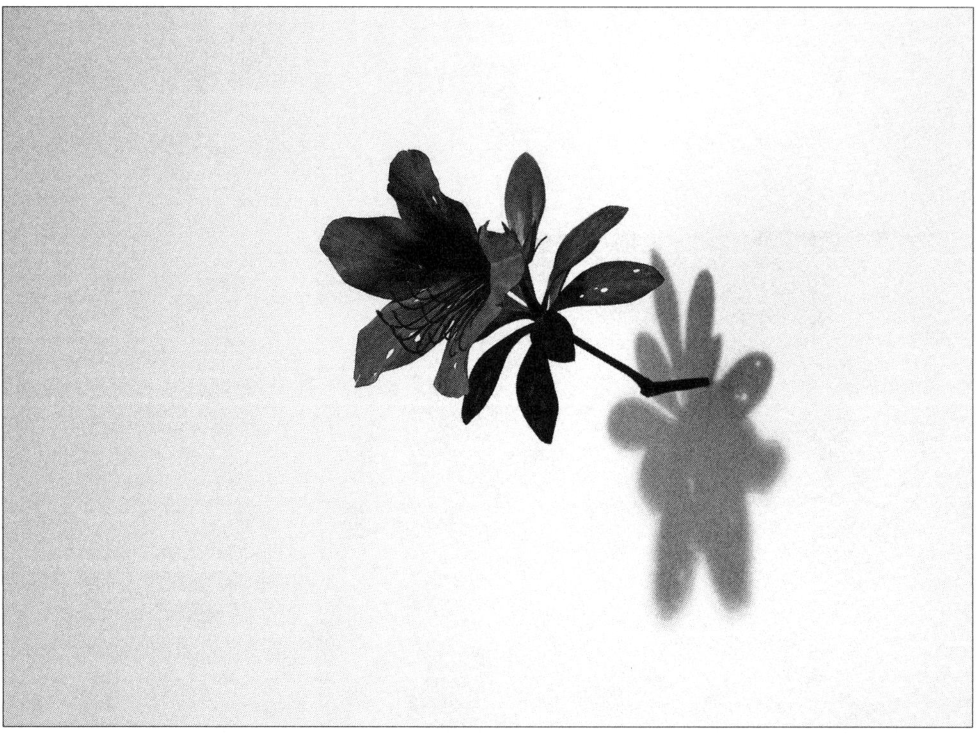

Figure 8.11
Yoshihiro Suda, *Azalea*, 2007. Paint on wood. © 2007 by Yoshihiro Suda. Courtesy of Gallery Koyanagi Tokyo.

glimpses into crucial sets of values and expectations surrounding 'the fate of the public'" (2010, 69).

Reflecting on public and counterpublic debates in relation to case studies of locative media art projects, Galloway considers how emergent and contingent forms of public are being both formed and informed by mobile media. As she suggests, these "messy" and "fluid" assemblages of mobile publics are not simply "networked," and scholars need to conduct more research into localized forms of mobile publics. Publics, rather than networks, provide a useful rubric for us to understand the converging and localized forms of social, locative, and mobile media practices with art, politics, and the environment in the Asia-Pacific. In keeping with Galloway's call for more work on the sociocultural and political uses of mobile media in constructing publics beyond the "private life–public space" trope, the examples we have discussed illustrate how social and intimate publics are beginning to be generated through art in the region.

Collectively, these examples show how, by making intimate connections in a public domain, mobile media are offering artists unique and powerful ways to become critically engaged in contemporary issues, in particular around environmental crisis and climate change.

Intimacy is something that enables intervention. However, the intervention it enables is not an external force to push against and resist. Rather—in tune with our wider argument for understanding climate change and the processes that contribute to it from *within*—the types of intimacy invoked through mobile media enable an involvement from within. In the next and concluding chapter, we bring together these points to argue for an approach to environment that addresses media, art, and climate change from within.

In this book, we have shown how intersections, connections, and disconnections are emerging between art, screen media, and climate change in the Asia-Pacific. This has offered us a particular prism through which to understand how experiences, critiques, and new forms of awareness around, and participation in relation to, climate change and environmental issues are emerging. As we have shown, these relationships are always subject to the contingencies of the places in which they emerge.

A central argument of this book is that we need to attend to the ways in which art and media are *part of* the issues and processes associated with climate change. Art and media are not simply texts, artifacts, and technologies that represent, reflect, and cause climate change. Instead, to understand how all three of these elements shape and constitute the environments we share with them, we need to appreciate how the processes they are respectively implicated in are entangled and, indeed, develop a series of ironic interdependencies. As Maxwell and Miller (2012) have argued, media are part of the "problem." Likewise the production of art—the biennials, the emergence of photo-sharing platforms like Instagram, and participatory digital art, as we discuss throughout the book, are inextricable from climate change.

Maxwell and Miller's conclusion to *Greening the Media* is highly convincing in offering readers a series of proposed "points of departure for greening the media," which are "green consumption, learning about media technology's eco-historical context, greening cultural labor and working conditions, pressing for greener governance, and cultivating forms of green citizenship rooted in eco-ethics and values of sustainability" (2012, 163). Clearly an agenda of greening the media would need to be situated as part of a wider sustainability process, yet because media are ubiquitous in everyday life and are part of the "technological sublime" that Maxwell and Miller problematize (163), media also emerge as a possible entry point to wider agendas for environmental sustainability. Indeed, as we have shown in chapters 4 and 8, mobile media and smartphones are being used by artists as technologies for generating intimate publics around climate change issues.

In the preceding chapters, we have explored contemporary and emerging screen ecologies in the Asia-Pacific, revealing the growing entanglements between media, art, and the environment in the region. In doing so, we have located new media as part of everyday life and as an essential element of how we understand the potential of public art to generate critical commentary and awareness of, and interventions in, climate change debates, and potentially also processes of change. We have argued that because mobile media and smartphones have the potential to generate "intimate publics" around issues generally, they subsequently enable new ways of constituting participatory and collaborative publics around climate change issues. We therefore see digital technologies emerging in ways that place them at the center of critical arts practice. However, these uses of media ultimately still carry the ironies that we highlighted at the beginning of the book: they depend on the very consumer cultures that have created a demand for smartphones and other digital technologies for their publics. The technological sublime, moreover, remains integral to the ways in which the demand for smartphones is constituted. This scenario is part of how mobile media are emerging as part of arts practice in the Asia-Pacific.

Art as a Critical Narrative in the Asia-Pacific

As we have shown in the preceding chapters, critical commentary on how we experience climate change and make environments is part of contemporary arts practice in the Asia-Pacific. In a global context where climate change is not simply a reality but also a debate, a critique, and an everyday anxiety, there exists a discursive domain wherein scholarship, arts practice, and activism come together as part of the wider environment in which critical publics are constituted. The visual and other nonverbal comments on this context that are made by artists have played a key role in the development of our discussion. We have interpreted the images that we have included here to some extent, but we also intend them to speak directly, in ways that go beyond words, to readers about the issues and experiences involved. The work of artists in this field makes the sublime disturbing, in ways that echo the agenda of Maxwell and Miller, as they likewise ask us to look beyond what we thought we could see, and attend to the usually ignored ways in which the technological sublime is implicated in climate change.

For instance, work like that of the aforementioned Indian artist Gigi Scaria brings to the fore ways of feeling about climate change that are hard to put into words. In Scaria's

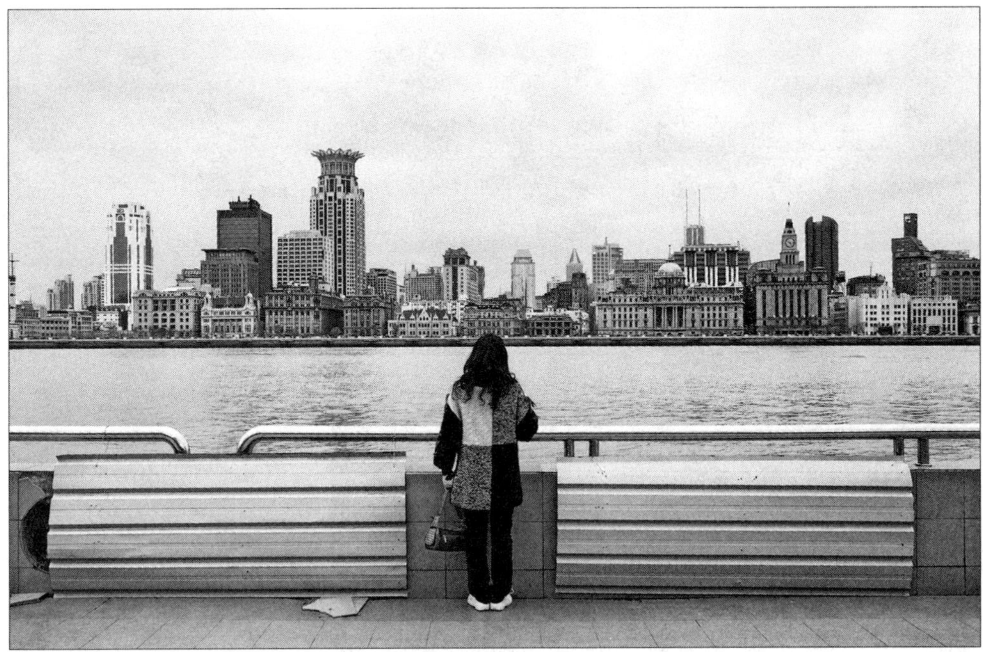

Figure 9.1
Gigi Scaria, *Conscience Keeper*, 2010. Courtesy of the artist.

Conscience Keeper (fig. 9.1), we watch a girl as she looks out at the Shanghai water and urban scape. There is a sense of distance, of inability to actually intervene or transgress beyond being a spectator. In another work, Scaria blurs New Delhi and Shanghai into a mirror composition. It is iconic and abstract, familiar and yet strange. Against the sublime shapes of the urban scape, the Yangtze River looks exploited and polluted.

Scaria's images speak of contemporary urbanity. They stand at the crossroads of media, art, and the environment, making powerful observations on contemporary everyday life in Asia that equally coincide with our focus on this very intersection. This is the space of the intimate stranger, whereby media are embedded within the fabric of the everyday. From *Conscience Keeper* to earlier works such as *Triviality of Everyday Existence* (2008), Scaria highlights the embedded mobility and immobility of contemporary life.

Ecology and environment are at the core of the artwork of the Australian artist Susan Norrie. In Australia, where climate change dominates political agendas and party policies, Norrie's images of disasters and disruption make it impossible to ignore the reality. In her 2012 video of 3/11 protests in Japan, Norrie captures the spirit of the protester

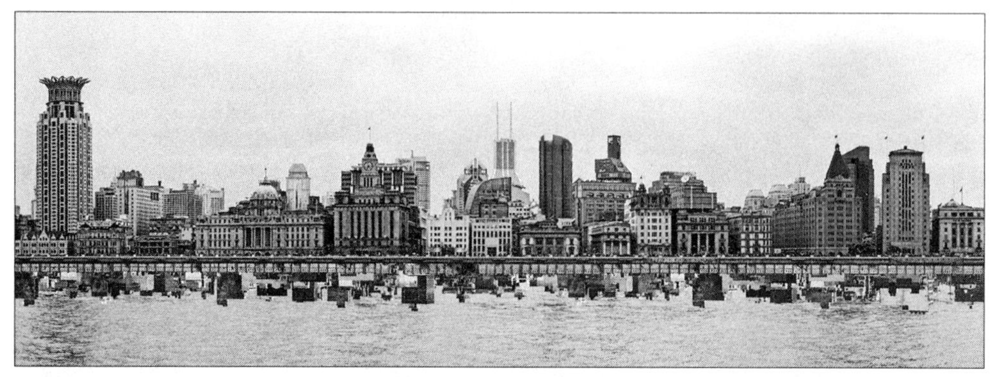

Figure 9.2
Gigi Scaria, *Equator*, 2011. Courtesy of the artist.

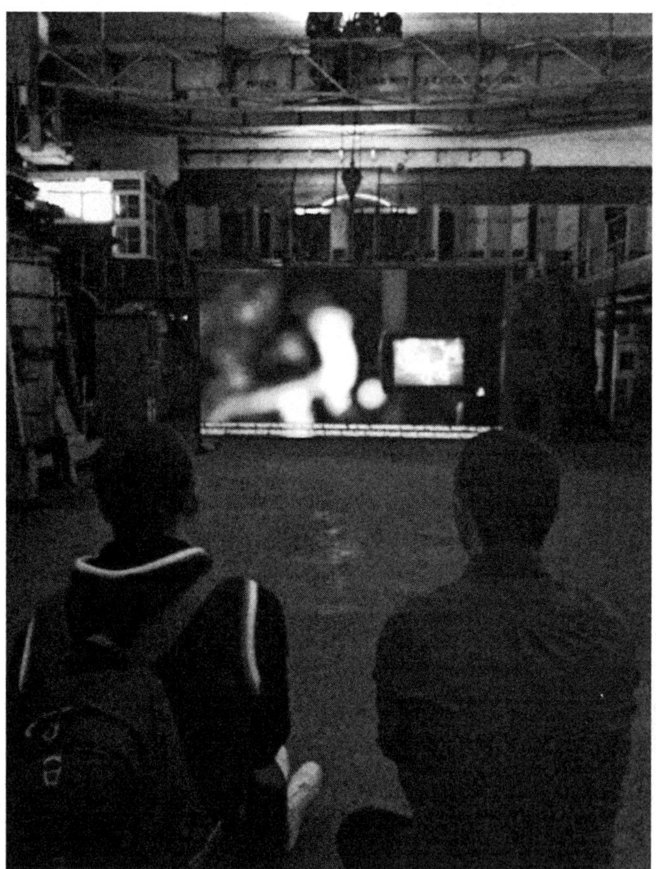

Figure 9.3
Susan Norrie, *Dissent*, Cockatoo Island power plant, 2014. Photo: Larissa Hjorth.

in the face of despair. For the Nineteenth Biennale of Sydney, she cleverly located the work within Cockatoo Island's disused power plant. It is not by accident that, in the last decade, Norrie has focused on the growing number of climate change disasters in the Asia-Pacific. The Australian art theorist Daniel Palmer argues:

The age of anthropogenic climate change coincides with the need for a more complex understanding of "nature," seriously complicating traditional formulations of the sublime. Instead of nature "out there," external to us—a modern conceit that has enabled all kinds of environmental disasters (in the name of progress)—global warming reveals a complete intermingling between the human and natural worlds. That we assume the planet will survive whatever eventuates, just not in a state fit for most animal or human habitation, could give rise to a paralyzing sublime. But a more radical eco-sublime is required to imagine new ways of feeling, alternative ways of thinking about our place within "nature," and new forms of collective empathy towards all manner of human and non-human others. (2014, 70)

These narratives of human and nonhuman, material and immaterial, and the relationships between media, art, and the environment are, for theorists such as Cubitt (2014), deeply interwoven with the politics and practices of new media regulation. For the Australian new media artist Lynette Wallworth, and especially in *Still: Waiting 2* (2006), media becomes the lens for heightening the relationship between "nature" and colonialism. With images of industrial wastelands contrasted with nature, the haunting by colonialism is palpable.

Another new media artist is Chris Howlett, who in *Metropolis: Part I–III* uses the machinima (machine cinema) made from a repurposed game (*SimCity Societies*) to play out some of the darkest scenarios of an urban dystopia with raging fires and mass destruction. Players and avatars are unable to see where the threat is coming from or where the next disaster will occur. In this game, the end is mass destruction, but Howlett seems to suggest that these game spaces are no longer too distant from real life. As we start to imagine how sustainable media futures might be played out, creative practice has a key role to play. Maxwell and Miller do not offer a definitive solution to the problem of greening the media but instead present a speculative fiction. When we bring together creative practice with the affordances of mobile media, new opportunities arise to consider how media art can participate in this context.

Mobile Entanglements

With the rise of personalized media (Bennett 2012), relationships between politics, intimacy, and publics are taking new forms that are amplified through the emergent

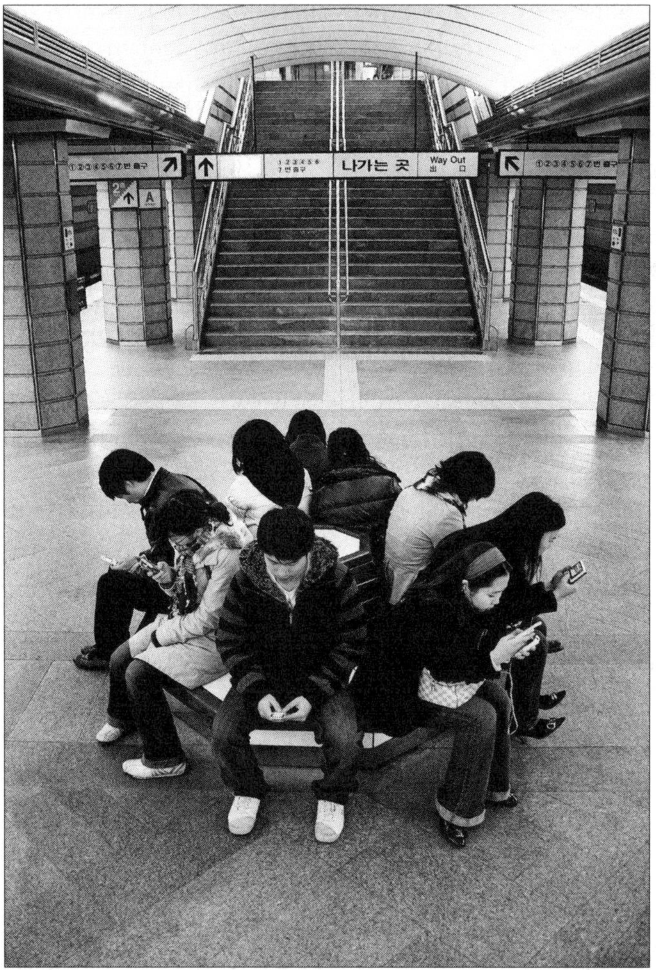

Figure 9.4
Gigi Scaria, *Triviality of Everyday Existence*, 2008. Courtesy of the artist.

forms of media and art practice we focus on in this book. Just as games have become an embedded part of everyday life (often through smartphones), they highlight the increasingly entangled role that new media are playing in our environment. One way in which we might understand these new entanglements is through what the media theorist Robert Grusin calls "premediation," whereby "the real is defined not in terms of representational accuracy, but in terms of liquidity or mobility" (2010, 3). As we have outlined in this book, various forms of mobility and immobility—across the technological, economic, political, and cultural—have seen screen culture, art, and the environment give way to new trajectories and cartographies.

As smartphones unevenly usher in new types of media practice and modes of copresence, divisions between digital and visual art blur. To understand the context of media today, it is not enough to examine the immaterial without understanding it as intertwined within the material. As Mimi Sheller (2014) argues in "Mobile Art: Out of Your Pocket" in reaction to Claire Bishop's dismissal of contemporary digital art:

I want to interrogate this digital disavowal not by re-visiting the recent nostalgic fascination with analog media, as Bishop does, but by re-situating contemporary mobile art as a far more diverse set of practices that are not simply about screening digital art on portable devices. Mobile art has in fact expanded the spatial and social field in which art takes place by experimenting with the mobile interface as a bridge between digital and physical space, a hybrid mediation of human sensory perception and technological connectivity. (2014, 376)

In keeping with Sheller's position, we have located art as embedded within contemporary media practice. By deploying highly pervasive metaphors like the platform (Gillespie 2010) and public and private "presence bleed" (Gregg 2011) to understand the contexts for the formation of intimate publics, we have not only endeavored to curate and contextualize intersections between art, media, and the environment as a nexus for understanding contemporary culture, but have also explored, through a series of case studies grounded in art, media, and environment, the ways in which practice today might be conceptualized and contextualized.

As media, art, and the environment are dynamic concepts, we have attempted to capture various snapshots of them intersecting at different points. Some chapters have focused on art, others on screen media, others again on the environment. Drawing from a variety of examples in the region, we have sketched some of the key debates, especially around changing notions of media practice, intimacy, and the public.

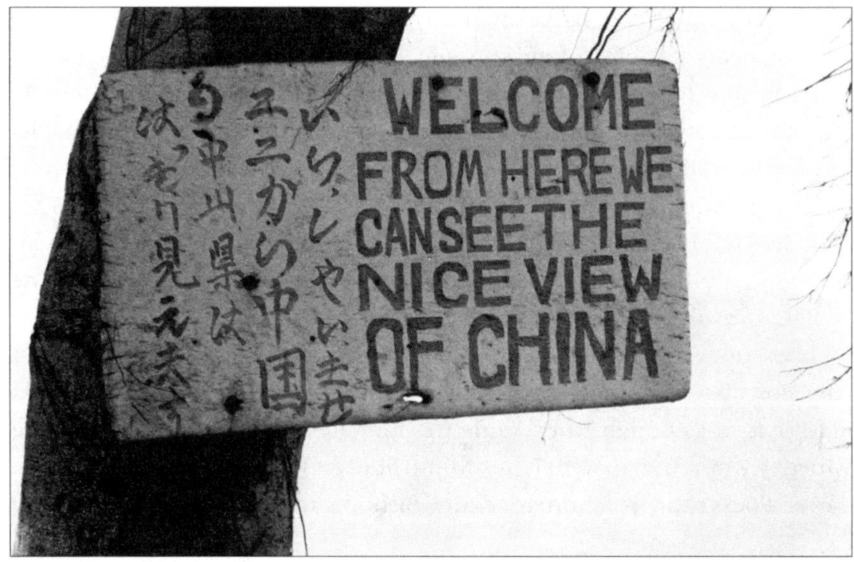

Figure 9.5
Asia Art Archive, from the publication *Mapping Asia* (2014). Courtesy of Monika Gilsing and Asia Art Archive, Hong Kong.

Rethinking the Region

Throughout the book, we have sought to explore some of the interrelations between the environment, screen culture, and regional localities. We have sought to reconcile two intersecting discussions that have often been left tacit, connecting the consumption and production of ICTs in the region with the politics of environmental pollution and emerging forms of activism. In particular, debates are only now beginning to acknowledge e-waste, and we need to more fully engage with it in new media and consumption areas.

Although the analysis of these kinds of relations in such a vast region can only ever be partial rather than comprehensive and definitive, we have suggested ways to rethink this emerging phenomenon. As we noted in chapter 2, the cartographies of the region can be understood as a series of maps—social, emotional, informational, geographic, technological, and political. In chapters 3 through 5, we contextualized the key inter-related concerns: the environment, screen culture, and regional localities. We then intensified our analysis by focusing on mobile and intimate (social) publics in the form of emergent sociocultural media practices, eco-cities, ecocriticism, and the role of the platform as a metaphor in and across art and media studies. Through these rubrics, we

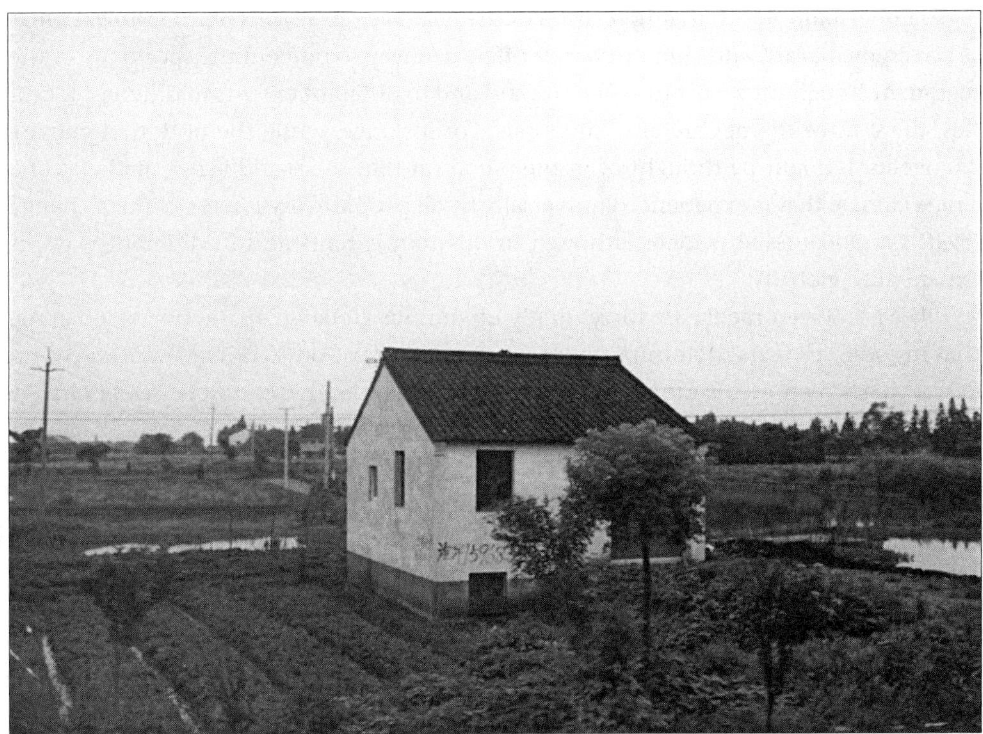

Figure 9.6
A mobile phone number tagged onto a home: a typical practice of locating a sense of home and
mobility in China. Photo: Larissa Hjorth.

argue that notions of the mobile, intimate, and public are being entangled in new
ways.

One of the most important theorists in the region, the aforementioned Kuan-Hsing
Chen, has moved toward understanding Asia's multiple histories in terms of what he
calls "Asia as method." In this revision of the region's contested identities consisting of
decolonization, deimperialization, and a Cold War hangover, Chen (2010) argues for
"Asian studies in Asia" to address the consequences of imperial histories. The notion of
the Asia-Pacific further complicates the region's multiple histories and violence, especially in the obvious financial differences between "developed" Asia and the "developing" Pacific. While discursive devices like the Asia Pacific Triennial of Contemporary
Art (APT) (discussed in detail in chap. 5) have sought to reconceptualize the region
through art practice, this has often left out the role of screen culture and the environment, despite these being two key features of the uneven region.

To understand the changing realities of the Asia-Pacific, a focus on the three themes of screen media, art, and climate change offers us a way to pull out the specificity of the region. Although they exist in different ways and to different extents in different localities, these three themes are also universals. For instance, while the notion of climate change itself might be thought of in the plural (in that the world is not made up of a single climate that is experienced universally by all people everywhere), climate change itself is a global issue, which, although to different extents and in different ways, is experienced globally.

Likewise, screen media are increasingly ubiquitous globally. Again, one could argue that in some contexts their importance is reduced in relation to others, but there is no doubt that screen media are embedded in, and indeed underpin, processes of change that are both locally and globally significant. It has therefore been highly significant to bring together these processes with a consideration of contemporary public art in the Asia-Pacific—art itself being a globally occurring phenomenon that is part of the environment, rather than simply a representation of it. Indeed, as we suggested earlier, the work of Scaria and other artists from the region should be understood not simply as visual representation but as part of the very critical debate that we are likewise engaging in by writing this book.

As we have shown in earlier chapters, the environmental debates that are emerging as specific to the region are part of contemporary arts practice as much as they are being generated through social media and other screen media practices. Therefore, to comprehend contemporary public art and the critical eco-aware publics that are emerging around it, we need to situate any analysis of public art in relation to the ways in which everyday lives are lived.

In the Asia-Pacific, this means engaging with questions about how digital and mobile screen media are used by ordinary people and how their uses of these media are being harnessed for new forms of public engagement. By focusing on the specific qualities and affordances of screen media and the ways that the possibilities for new forms of intimacy they offer are generated in specifically local Asia-Pacific contexts, we have suggested that we are now in a context where the role of public art in generating critical encounters with climate change goes beyond representation. Instead it involves new layers of technologically mediated experience and action, accessible both locally and globally, and therefore shapes new ways of engaging with both art and environmental issues.

It is these questions that we urge readers to look to when considering how contemporary art both relates to climate change and might be engaged in action against climate change. Both exist in complex relations to each other, yet these relations cannot be understood, or harnessed to work toward environmental sustainability, without a recognition of how screen ecologies and digital mobile technologies play a dual role, as they interface with everyday life, art, and the environment, while also making, and being part of, that very environment.

A Way Forward?

There is no ready solution to this problem. However, the example of the Asia-Pacific invites us to consider, if arts practice is able both to draw on the ways in which mobile media and apps create "intimate publics" and to harness them to confront issues around climate change, what the next step for art in working toward a sustainable future might be.

We end the book with a set of propositions. These are not solutions. Rather, they stand as an invitation to scholars and artists to consider how a refigured media arts practice might take steps toward an environmentally sustainable future.

1. Refigure the technological sublime to engage not only with the consumption and production of ICTs in the region but also with the politics of environmental pollution, especially e-waste. In particular, we need to stop thinking about North and South binaries, whereby the South is viewed as the dumping grounds for Northern e-consumption. Artists can provide ways in which to intervene in these politics and suggest a more ambivalent and nuanced understanding.

2. Engage with recycled and older media in ways that do not require infrastructures that are technologically and environmentally costly. We can create media art that is based on existing media practices and uses that do not generate new data or require new technologies. This involves putting into practice media uses that do not always require new smartphones, do not need to produce new images and e-waste, and are technologically and digitally thrifty.

3. Focus on the social, and particularly on forms of intimacy, as ways to develop green media art. Through participatory hacking of everyday mobile media practices through art, along with playfulness, we propose to engage intimate publics to recognize and refigure their media practices in a more knowing and sustainable manner. We argue that artists are providing new ways for thinking about being green—what we call "green media art"—that expand on discussions within media studies.

4. Offer models for understanding the complex ways in which ICTs are implicated in material and immaterial ways in everyday life, and how these can shape how we reflect on and use our everyday media. The examples we have provided show a new

media agility being pioneered by artists as they move in and around existing digital, mobile, and social media. These emerging art forms are redefining what it means to practice activism through collaborative platforms, noncommercial software, and so on, in ways that are environmentally sustainable.

Interventions that seek to address these issues signify how and where arts practice might begin to contribute to making a democratic and environmentally sustainable world. Yet they will need to be coupled with technological interventions and political will, as well as with publics. Engaging with contemporary everyday media such as camera phone apps is not only a democratization of media but also a playful way for framing interactions between artists and amateurs.

References

Amin, Ash. 2008. Collective culture and urban public space. *City* 2 (1): 5–24.

Anderson, Alison. 1997. *Media, Culture, and the Environment*. New York: Routledge.

Anderson, Alison. 2014. *Media, Environment, and the Network Society*. London: Palgrave Macmillan.

Andrew, Brooke, Thomas J. Berghuis, Lisa Havilah, and Aaron Seeto. 2010. *Edge of Elsewhere*. Campbelltown/Sydney: Campbelltown Arts Centre, Gallery 4A.

Ang, Ien. 2003. Cultural translation in a globalised world. In *Complex Entanglements: Art, Globalisation, and Cultural Difference*, ed. Nikos Papastergiadis, 30–41. London: Rivers Oram Press.

Antoinette, Michelle. 2007. Deterritorializing aesthetics: International art and its new cosmopolitanisms, from an Indonesian perspective. In *Cosmopatriots: On Distant Belongings and Close Encounters*, ed. Edwin Jurriens and Jeroen de Kloet, 205–234. New York: Rodopi.

Antoinette, Michelle, and Caroline Turner, eds. 2014. *Contemporary Asian Art and Exhibitions: Connectivities and World-making*. Canberra: ANU Press.

Appadurai, Arjun. 1996. *Modernity at Large: Cultural Dimensions of Globalization*. Minneapolis: University of Minnesota Press.

Arnold, Michael, Craig Bellamy, Brent Coker, Martin Gibbs, Paul Hill, and Bjorn Nansen. 2014. *Framing the NBN: Public Perceptions and Media Representations*. Melbourne: Institute for a Broadband Enabled Society.

Arrighi, Giovanni. 1994. *The Long Twentieth Century: Money, Power, and the Origins of Our Times*. New York: Verso.

Arrighi, Giovanni. 2005. States, markets, and capitalism, east and west. In *Semináro Internacional REG GEN: Alternativas Globalização (8 al 13 de Octubre de 2005, Hotel Gloria, Rio de Janeiro, Brasil)*. Rio de Janeiro: UNESCO, Organización de las Naciones Unidas para la Educación, la Ciencia y la Cultura. http://bibliotecavirtual.clacso.org.ar/ar/libros/reggen/pp25.pdf.

Arrighi, Giovanni. 2009. *Adam Smith in Beijing: Lineages of the 21st Century*. New York: Verso.

Arrighi, Giovanni, Takashi Hamashita, and Mark Selden. 2003. *The Resurgence of East Asia: 500, 150 and 50 Year Perspectives*. London: Routledge.

Artist as Family. 2014. http://theartistasfamily.blogspot.com.au.

Ashford, Doug, Wendy Ewald, Nina Felshin, and Patricia C. Philipps. 2006. A conversation on social collaboration. *Art Journal* 65 (2): 5–83.

Asia Art Archive. 2011a. *The And: An Expanded Questionnaire on the Contemporary*. http://www.aaa.org.hk/FieldNotes/Details/1167 (accessed February 24, 2014).

Asia Art Archive. 2011b. Talk Series Residency Programme. Presentations by Young-Hae Chang Heavy Industries. http://www.aaa.org.hk/Programme/Details/12 (accessed February 24, 2014).

Asian Development Bank. 2012. *Green Urbanization in Asia: Key Indicators for Asia and the Pacific 2012: Special Chapter*. http://www.adb.org/sites/default/files/publication/29940/ki2012-special-chapter.pdf.

Associated Press. 2014. Gao Yu arrested by Chinese authorities. *Guardian*, May 8. http://www.theguardian.com/world/2014/may/08/gao-yu-arrested-by-chinese-authorities.

Augé, Marc. 2013. Architecture and non-places. Lecture given at Melbourne University, December 13, 2013.

Barnett, Jon, and W. Neil Adger. 2003. Climate dangers and atoll countries. *Climatic Change* 61 (3): 321–337.

BBC Editorial. 2014. Prisoner in China's plea for help found in Saks bag. *BBC News*, May 1, 2014. http://www.bbc.com/news/world-us-canada-27237870.

Beaulieu, Anne. 2010. Research note: From co-location to co-presence; Shifts in the use of ethnography for the study of knowledge. *Social Studies of Science* 40 (3): 453–470.

Beeson, Mark. 2013. Is this the end of the "Asian century"? *Conversation*, October 29, 2013. http://theconversation.com/is-this-the-end-of-the-asian-century-19616.

Behrens, Erik, Franziska U. Schwarzkopf, Joke F. Lübbecke, and Claus W. Böning. 2012. Model simulations on the long-term dispersal of 137Cs released into the Pacific Ocean off Fukushima. *Environmental Research Letters* 7.

Belting, Hans. 2009. *The Global Art World: Audiences, Markets and Museums*. Ostfildern, Germany: Hatje Cantz.

Belting, Hans, Jacob Birken, Andrea Buddensieg, and Peter Weibel. 2011. *Global Studies: Mapping Contemporary Art and Culture*. Ostfildern, Germany: Hatje Cantz.

Bennett, Jane. 2010. *Vibrant Matter: A Political Economy of Things*. Durham: Duke University Press.

Bennett, W. Lance. 2012. The personalization of politics, political identity, social media, and changing patterns of participation. *Annals of the American Academy of Political and Social Science* 644 (1): 20–39.

Berlant, Lauren. 1998. Intimacy: A special issue. *Critical Inquiry* 24 (2): 281–288.

Berry, Chris. 2013. Shanghai's public screen culture: Local and coeval. In *Public Space, Media Space*, ed. Chris Berry, Janet Harbord, and Rachel Moore, 110–114. New York: Palgrave Macmillan.

Berry, Chris, Nicola Liscutin, and Jonathan D. Mackintosh. 2009. *Cultural Studies and Cultural Industries in Northeast Asia: What a Difference a Region Makes*. Hong Kong: Hong Kong University Press.

Biennial Foundation. n.d. http://www.biennialfoundation.org/biennial-map (accessed February 24, 2014).

Bijker, Wiebe E., Thomas Parke Hughes, and Trevor Pinch. 1987. *The Social Construction of Technological Systems: New Directions in the Sociology and History of Technology*. Cambridge, MA: MIT Press.

Bishop, Claire. 2004. Antagonism and relational aesthetics. *October* 110 (fall): 51–79.

Bishop, Claire. 2006a. The social turn: Collaboration and its discontents. *Artforum*, February, 179–185.

Bishop, Claire. 2006b. Reply. *Artforum*, May, 22–23.

Bishop, Claire. 2006c. *Participation*. London: Whitechapel.

Bishop, Claire. 2012a. Digital divide: Contemporary art and new media. *Artforum*, September, http://artforum.com/talkback/id=70724.

Bishop, Claire. 2012b. *Artificial Hells: Participatory Art and the Politics of Spectatorship*. London: Verso.

Bolter, Jay David, and Richard Grusin. 1999. *Remediation: Understanding New Media*. Cambridge, MA: MIT Press.

Bourriaud, Nicolas. 2002. *Relational Aesthetics*. Dijon: Les Presses du Réel. First published in French, 1998.

Braddock, Alan, and Christoph Irmscher. 2009. *A Keener Perception: Ecocritical Studies in American Art History*. Tuscaloosa: University of Alabama Press.

Bristow, Frances. 2011. The impact of dams on the Sainte-Marguerite and Romaine rivers on the phytoplankton communities and the physical-chemical properties of their estuaries. http://www.archipel.uqam.ca/4197.

Brown, Andrew. 2014. *Art and Ecology Now*. London: Thames & Hudson.

Bruns, Axel. 2005. Some exploratory notes on produsers and produsage. http://distributedcreativity.typepad.com/idc_texts/2005/11/some_explorator.html (accessed April 25, 2014).

Bruns, Axel. 2008. The future is user-led: The path towards widespread produsage. *Fibreculture Journal* 11.

Burgess, Jean. 2007. Vernacular creativity and new media. PhD thesis, Queensland University of Technology.

Burgess, Jean, and Joshua Green. 2009. *YouTube: Online Video and Participatory Culture*. Cambridge: Polity.

Burgmann, Verity, and Hans Baer. 2012. *Climate Politics and the Climate Movement in Australia*. Melbourne: Melbourne University Press.

Butt, Zoe. 2014. Red tape and digital talismans: Shaping knowledge beneath surveillance. In *Art in the Asia-Pacific: Intimate Publics*, ed. Larissa Hjorth, Natalie King, and Mami Kataoka, 91–104. New York: Routledge.

Carbon Arts. 2014. *Curating Cities*. http://www.carbonarts.org/projects/curating-cities (accessed February 17, 2015).

Carruth, Allison, and Robert P. Marzec. 2014. Environmental visualization in the anthropocene: Technologies, aesthetics, ethics. *Public Culture* 26 (2): 205–212.

Castells, Manuel. 1996. *The Rise of the Networked Society*. New York: John Wiley & Sons.

Castells, Manuel. 1997. *The Power of Identity*, vol. 2: *The Information Age: Economy, Society and Culture*. London: Blackwell.

Chellaney, Brahma. 2012. Asia's worsening water crisis. *Survival* 54 (2): 143–156.

Chellaney, Brahma. 2013a. *Water, Peace, and War: Confronting the Global Water Crisis*. London: Rowman & Littlefield.

Chellaney, Brahma. 2013b. Why water is becoming the new oil. *Project Syndicate*, August 7. http://www.project-syndicate.org/commentary/why-water-is-becoming-the-new-oil-by-brahma-chellaney (accessed November 27, 2013).

Chen, Gang. 2010. *Towards a Livable and Sustainable Urban Environment: Eco-cities in East Asia*. Singapore: World Scientific.

Chen, Kuan-Hsing. 2010. *Asia as Method: Toward Deimperialization*. Durham: Duke University Press.

Chesher, Chris. 2004. Neither gaze nor glance, but glaze: Relating to console game screens. *SCAN: Journal of Media Arts Culture* 1 (1). http://scan.net.au/scan/journal (accessed February 10, 2007).

Chim↑Pom. 2011a. LEVEL7 feat. Myth of Tomorrow / Chim↑Pom. https://www.youtube.com/watch?v=bKZBhbzLqv4 (accessed April 17, 2014).

Chim↑Pom. 2011b. *Real Times*. http://chimpom.jp/realtimes.html (accessed April 17, 2014).

Ching, Leo T. S. 2000. Globalizing the regional, regionalizing the global: Mass culture and Asianism in the age of late capital. *Public Culture* 12 (1): 233–257.

Chiu, Melissa, and Benjamin Genocchio. 2010. *Contemporary Asian Art*. London: Thames & Hudson.

Chiu, Melissa, and Benjamin Genocchio. 2011. *Contemporary Art in Asia: A Critical Reader*. Cambridge, MA: MIT Press.

Chow, Rey. 1997. Listening otherwise, music miniaturized: A different type of question about revolution. In *Doing Cultural Studies: The Story of the Sony Walkman*, ed. Paul du Gay et al., 135–140. London: Open University Press.

Christensen, Christian. 2011. Twitter revolutions? Addressing social media and dissent. *Communication Review* 14 (3): 155–157.

Chua, Beng-Huat. 2000. *Consumption in Asia*. London: Routledge.

Chung, Peichi. 2016. The globalization of game art in Southeast Asia. In *The Routledge Handbook to New Media in Asia*, ed. Larissa Hjorth and Olivia Khoo, 402–415. New York: Routledge.

Citizen Lab. 2013. *An Overview of Indonesian Internet Infrastructure and Governance*. https://citizenlab.org/2013/10/igf-2013-an-overview-of-indonesian-internet-infrastructure-and-governance (accessed May 4, 2014).

Clark, John. 1998. *Modern Asian Art*. Sydney: Craftsman House/G+B Arts International.

Clark, John. 2007. Histories of the Asian "new": Biennales and contemporary art. In *Asian Art History in the Twenty-first Century*, ed. Vishakha N. Desai, 229–249. New Haven: Yale University Press.

Clark, John. 2010. Biennales as structures for the writing of art history: The Sian perspective. In *The Biennale Reader*, ed. Elena Filipovic, Marieke van Hal, and Solveig Ovstebo, 164–183. Ostfildern, Germany: Hatje Cantz.

Climarte. 2015. http://climarte.org/about/. Accessed 10 October 2015.

Couldry, Nick. 2012. *Media, Society, World: Social Theory and Digital Media Practice*. London: Polity Press.

Couldry, Nick, and Anna McCarthy. 2004. *MediaSpace: Place, Scale, and Culture in a Media Age*. London: Routledge.

Crawford, Holly. 2008. *Artistic Bedfellows: Histories, Theories, and Conversations in Collaborative Art Practices*. Lanham, MD: University Press of America.

Cruickshank, Alan. 2006. Interview: Charles Merewether. *Broadsheet* 35 (2). http://www.cacsa.org.au//Wordpress/yoo_bigeasy_demo_package_wp/wp-content/uploads/Broadsheet/2006/35_2/merewether.pdf (accessed May 28, 2014).

CSIRO. 2006. *CSIRO Sustainable Ecosystems Sustainability Report 2006–07*. http://www.clw.csiro.au/publications/.

Cubitt, Sean. 2005. *Ecomedia*. Amsterdam: Rodopi.

Cubitt, Sean. 2014. Regional standardization. In *Art in the Asia-Pacific: Intimate Publics*, ed. Larissa Hjorth, Natalie King, and Mami Kataoka, 134–145. New York: Routledge.

Cullen, Simon. 2013. Abbot backs plans to build more dams. *ABC News*, February 14.

Curtin, Michael. 2003. Media capital: Towards the study of spatial flows. *International Journal of Cultural Studies* 6 (2): 202–227.

Dawson, Grant, and Sonia Farber. 2012. *Forcible Displacement throughout the Ages: Towards an International Convention for the Prevention and Punishment of the Crime of Forcible Displacement.* Boston: Martinus Nijhoff.

Deleuze, Gilles, and Félix Guattari. 1987. *A Thousand Plateaus: Capitalism and Schizophrenia.* Trans. Brian Massumi. Minneapolis: University of Minnesota Press.

Derrida, Jacques. 1982. *Différance.* Trans. Alan Bass, 3–27. Chicago: University of Chicago Press.

Diamond, Larry, and Marc Plattner. 2012. *Liberation Technology: Social Media and the Struggle for Democracy.* Baltimore: The John Hopkins University Press.

Dirlik, Arif. 2005. Asia Pacific studies in an age of global modernity. *Inter-Asia Cultural Studies* 6 (2): 158–170.

Dirlik, Arif. 2007. Global South: Predicament and promise. *Global Society* 1 (1): 12–23.

Dourish, Paul. 2005. The culture of information: Ubiquitous computing and representations of reality. *Designing Ubiquitous Information Environments: Socio-Technical Issues and Challenges IFIP— The International Federation for Information Processing* 185: 23–26.

Downey, Anthony. 2009. An ethics of engagement: Collaborative art practices and the return of the ethnographer. *Third Text* 23 (5): 593–603.

Eckersley, Robyn. 2004. *The Green State.* Cambridge, MA: MIT Press.

Elkins, James. 2007. *Is Art History Global?* New York: Routledge.

Elliott, Anthony, and John Urry. 2010. *Mobile Lives.* London: Routledge.

Elsner, James B., James P. Kossin, and Thomas H. Jagger. 2008. The increasing intensity of the strongest tropical cyclones. *Nature* 455:92–95.

Environmental Research Initiative for Art (ERIA). 2014. http://www.niea.unsw.edu.au/research/ organisations/environmental-research-initiative-art-eria (accessed February 17, 2015).

Enwezor, Okwui. 2002. *Mega Exhibitions: Antinomies of a Transnational Global Form.* Munich: Wilhelm Fink.

Enwezor, Okwui. 2009. The postcolonial constellation: Contemporary art in a state of permanent transition. In *Antinomies of Art and Culture: Modernity, Postmodernity, and Contemporaneity*, ed. Terry Smith, Okwui Enwezor, and Nancy Condee, 207–234. Durham: Duke University Press.

Eriksen, Thomas Hylland. 2009. Living in an overheated world: Otherness as a universal condition. *Acta Historica Universitatis Klaipedensis* 19.

Eriksen, Thomas Hylland. 2012. *An Overheated World.* http://www.sv.uio.no/sai/english/research/projects/overheating/publications/overheated-world.pdf (accessed May 28, 2014).

Estok, Simon, and Won-Chung Kim. 2013. *East Asian Ecocriticisms: A Critical Reader.* New York: Palgrave Macmillan.

Farbotko, Carol, and Heather Lazrus. 2012. The first climate refugees? Contesting global narratives of climate change in Tuvalu. *Global Environmental Change* 22:382–390.

Farman, Jason. 2010. Mapping the digital empire: Google Earth and the process of postmodern cartography. *New Media and Society* 12:869–888.

Farman, Jason. 2012. *Mobile Interface Theory: Embodied Space and Locative Media.* Oxford: Routledge.

Filipovic, Elena, Marieke van Hal, and Solveig Ovstebo. 2010. *The Biennial Reader.* Bergen: Bergen Kunsthall; Ostfildern: Hatje Cantz.

Fortunati, Leopoldina. 2002. Italy: Stereotypes, true and false. In *Perpetual Contact: Mobile Communications, Private Talk, Public Performance,* ed. James E. Katz and Mark Aakhus, 42–62. Cambridge: Cambridge University Press.

Foster, Hal. 1994. The artist as ethnographer. In *Global Visions: Toward a New Internationalism in the Visual Arts,* ed. Jean Fisher, 12–19. London: Kala Press.

Frieling, Rudolf, Robert Atkins, Boris Groys, and Lev Manovich. 2008. *The Art of Participation: 1950 to Now.* London: Thames & Hudson.

Fu, Bo-Jie, et al. 2010. Three Gorges Project: Efforts and challenges for the environment. *Progress in Physical Geography* 34 (6): 741–754.

Galloway, Anne. 2010. Mobile publics and issues-based art and design. In *The Wireless Spectrum,* ed. Barbara Crow, Michael Longford, and Kim Sawchuk, 63–76. Toronto: University of Toronto Press.

Gardner, Anthony, and Charles Green. 2014. Mega-exhibitions, new publics and Asian art biennials. In *Art in the Asia-Pacific: Intimate Publics,* ed. Larissa Hjorth, Natalie King, and Mami Kataoka, 23–36. New York: Routledge.

Giddens, Anthony. 1992. *The Transformation of Intimacy.* Stanford, CA: Stanford University Press.

Gillespie, Tarleton L. 2010. The politics of "platforms." *New Media and Society* 12 (3): 347–364.

Goggin, Gerard. 2011. *Global Mobile Media.* London: Routledge.

Gómez Cruz, Edgar, and Elisenda Ardèvol. 2013. Some ethnographic notes on a Flickr group. *Photographies* 6 (1): 35–44.

Green, Charles. 2001. *The Third Hand: Collaboration in Art from Conceptualism to Postmodernism*. Sydney: University of New South Wales Press.

Greene, Rachel. 2004. *Internet Art*. London: Thames & Hudson.

Gregg, Melissa. 2011. *Work's Intimacy*. London: Polity.

Grusin, Robert. 2010. *Premediation: Affect and Mediality after 9/11*. New York: Palgrave Macmillan.

Guo, Yingjie. 2004. *Cultural Nationalism in Contemporary China: The Search for National Identity under Reform*. New York: Routledge.

Gwangju Biennale. 2012. *Program*. http://www.gwangjubiennale.org/eng/gb/past/program/?no=9 &mode=view&bn_idx=62 (accessed January 15, 2014).

Hajdu, Sue. 2009. Missing in the Mekong. *Contemporary Visual Art and Culture Broadsheet* 38 (4): 267–269.

Hall, Stuart. 1997. *Representation: Cultural Representations and Signifying Practices*. London: Sage.

Hannam, Kevin, Mimi Sheller, and John Urry. 2006. Editorial: Mobilities, immobilities, and moorings. *Mobilities* 1 (1): 1–22.

Hanru, Hou, and Natalie King. 2013. Machines of knowledge and experimentation. *Art Monthly*. http://asialink.unimelb.edu.au/__data/assets/pdf_file/0009/749223/King_and_Hanru.pdf (accessed January 15, 2014).

Harris, Jonathan. 2011. *Globalization and Contemporary Art*. Malden, MA: Wiley-Blackwell.

Hattam, Jennifer. 2011. Traditional landscapes with a twist: Photographer Yao Lu makes mountains out of China's rubble heaps. *Treehugger*. http://www.treehugger.com/culture/traditional -landscapes-with-a-twist-photographer-yao-lu-makes-mountains-out-of-chinas-rubble-heaps .html.

Hawkins, Gay. 2005. *The Ethics of Waste: How We Relate to Rubbish*. Lanham, MD: Rowman & Littlefield.

Hawkins, Gay. 2011. Packaging water: Plastic bottles as market and public devices. *Economy and Society* 40 (4): 534–552.

Hawkins, Gay, Emily Potter, and Kane Race. 2015. *Plastic Water: The Social and Material Life of Bottled Water*. Cambridge, MA: MIT Press.

Hjorth, Larissa. 2005. Locating mobility: Practices of co-presence and the persistence of the postal metaphor in SMS/MMS mobile phone customization in Melbourne. *Fibreculture Journal* 6.

Hjorth, Larissa. 2007. Snapshots of almost contact. *Continuum* 21 (2): 227–238.

Hjorth, Larissa. 2008a. Guest editor's introduction. *Games and Culture* 3 (January): 3–12.

Hjorth, Larissa. 2008b. The game of being mobile: One media history of gaming and mobile technologies in Asia-Pacific. *Convergence: The International Journal of Research into New Media Technologies* 13 (4): 369–381.

Hjorth, Larissa. 2009. *Mobile Media in the Asia-Pacific: Gender and the Art of Being Mobile.* London: Routledge.

Hjorth, Larissa. 2010. The politics of being mobile: A case study of a different model for conceptualizing mobility, gaming and play. In *Digital Cityscapes,* ed. Adriana de Souza e Silva and Daniel Sutko, 83–99. New York: Peter Lang Group.

Hjorth, Larissa. 2013. Frames of discontent: Social media, mobile intimacy, and the boundaries of media practice. In *New Visualities, New Technologies: The New Ecstasy of Communication,* ed. Hille Koskela and John Macgregor Wise, 99–118. New York: Ashgate.

Hjorth, Larissa, and Michael Arnold. 2013. *Online@AsiaPacific.* New York: Routledge.

Hjorth, Larissa, and Yonnie Kim. 2011. The mourning after: A case study of social media in the 3.11 earthquake disaster in Japan. *Television and New Media* 12 (6): 552–559.

Hjorth, Larissa, Natalie King, and Mami Kataoka, eds. 2014. *Art in the Asia-Pacific: Intimate Publics,* New York: Routledge.

Hjorth, Larissa, and Sun Sun Lim. 2012. Mobile intimacy in an age of affective mobile media. *Feminist Media Studies,* December, 1–8.

Hjorth, Larissa, and Sarah Pink. 2014. New visualities and the digital wayfarer: Reconceptualizing camera phone photography and locative media. *Mobile Media and Communication* 2:40–57.

Hjorth, Larissa, and Kristen Sharp. 2014. The art of ethnography: The aesthetics or ethics of participation? *Visual Studies Journal* 29 (2): 128–135.

Hochman, Nadav, and Lev Manovich. 2013. Zooming into an Instagram city. *First Monday* 18 (7). http://firstmonday.org/ojs/index.php/fm/article/view/4711/3698 (accessed February 17, 2015).

Hochschild, Arlie. 1983. *The Managed Heart: Commercialization of Human Feeling.* Berkeley: University of California Press.

Hochschild, Arlie. 2012. *The Outsourced Self.* New York: Metropolitan Press.

Hoekstra, Arjen, and Ashok Chapagain. 2008. *Globalization of Water: Sharing the Planet's Freshwater Resources.* Oxford: Blackwell.

Hofman, Florentijn. 2013. *Projects.* http://www.florentijnhofman.nl/dev/project.php?id=192 (accessed January 15, 2013).

Hong Kong Economic and Trade Office, New York. 2013. Director highlights Hong Kong's growing arts and culture. http://www.hketony.gov.hk/ny/activities/11/director_highlights.htm (accessed February 24, 2014).

Horst, Heather. 2013. The infrastructures of mobile media: Towards a future research agenda. *Mobile Media and Communication* 1 (1): 147–152.

Hudson, Chris. 2014. Green is the new green: Eco-aesthetics in Singapore. In *Green Consumption: The Global Rise of Eco-Chic*, ed. Bart Barendregt and Rivke Jaffa, 86–99. London: Bloomsbury.

Huhtamo, Erkki, and Jussi Parikka, eds. 2011. *Media Archaeology: Approaches, Applications, and Implications*. Berkeley: University of California Press.

Hujatnikajennong, Agung. 2013. Jogya Biennale XII 2013 Equator #2: Curatorial Concept. http://www.biennalejogja.org/2013/about/jogja-biennale-xii-equator-2/?lang=en (accessed February 24, 2014).

Hvistendahl, Mara. 2008. China's Three Gorges Dam: An environmental catastrophe? *Scientific American.* http://www.scientificamerican.com/article/chinas-three-gorges-dam-disaster (accessed January 15, 2013).

Ikeda, I. 2008. Five floating isles. http://www.ikedawater.org/2008Kedogawa/2008KedogawaEng.html (accessed November 11, 2015).

Illouz, Eva. 2007. *Cold Intimacies: The Making of Emotional Capitalism*. Cambridge: Polity Press.

Ingold, Tim. 2007. *A Brief History*. London: Routledge.

Ingold, Tim. 2008. Bindings against boundaries: Entanglements of life in an open world. *Environment and Planning A* 40:1769–1810.

Ingold, Tim. 2010. Ways of mind-walking: Reading, writing, painting. *Visual Studies* 25 (1): 15–23.

Ingold, Tim. 2011. Reply to David Howes. *Social Anthropology* 19 (3): 323–327.

Ingold, Tim. 2013. *Making*. Oxford: Routledge.

Instagram. 2013. *Top Locations on Instagram in 2013*. http://blog.instagram.com/post/69877035043/top-locations-2013 (accessed February 17, 2015).

IPCC. 2011. *Managing the Risks of Extreme Events and Disasters to Advance Climate Change Adaptation (SREX)*. http://www.ipcc-wg2.gov/SREX (accessed February 17, 2015).

IPCC. 2013. *Fifth Assessment Report*. https://www.ipcc.ch/report/ar5/wg1 (accessed February 17, 2015).

Ito, Mizuko. 2002. Mobiles and the appropriation of place. *Receiver 8.*

Ito, Mizuko, and Daisuke Okabe. 2005. Intimate visual co-presence. Paper presented at Ubicomp, Takanawa Prince Hotel, Tokyo, Japan, September 11–14. http://www.itofisher.com/mito (accessed February 17, 2015).

Ito, Mizuko, Daisuke Okabe, and Misa Matsuda, eds. 2005. *Personal, Portable, Pedestrian: Mobile Phones in Japanese Life*. Cambridge, MA: MIT Press.

Iwabuchi, Koichi, and Beng-Huat Chua, eds. 2009. *East Asian Pop Culture: Analysing the Korean Wave*. Hong Kong: Hong Kong University Press.

Jamieson, Lyn. 1998. *Intimacy: Personal Relationships in Modern Societies*. Cambridge: Polity Press.

Jenkins, Henry. 2006. *Convergence Culture: Where Old and New Media Collide*. New York: NYU Press.

Jolly, Margaret. 2007. Imagining Oceania: Indigenous and foreign representations of a sea of islands. *Contemporary Pacific* 19 (2): 508–545.

Jurgenson, Nathan. 2011. The faux-vintage photo. *The Society Pages*. http://thesocietypages.org/cyborgology/2011/05/14/the-faux-vintage-photo-full-essay-parts-i-ii-and-iii/.

Jurriens, Edwin. 2014. Mediating the metropolis: New media art as a laboratory for urban ecology in Indonesia. In *Art in the Asia-Pacific: Intimate Publics*, ed. Larissa Hjorth, Natalie King, and Mami Kataoka, 173–190. London: Routledge.

Kaiman, Jonathan. 2013. Tiananmen Square online searches censored by Chinese authorities. *BBC Online*, June 4. http://www.theguardian.com/world/2013/jun/04/tiananmen-square-online-search-censored (accessed February 17, 2015).

Kataoka, Mami. 2014. Intimacy and distance with the public. In *Art in the Asia-Pacific: Intimate Public*, ed. Larissa Hjorth, Natalie King, and Mami Kataoka, 123–132. New York: Routledge.

Kenning, Dean. 2009. Art relations and the presence of absence. *Third Text* 23 (4): 435–446.

Kester, Grant. 2004. *Conversation Pieces: Community and Communication in Modern Art*. Berkeley: University of California Press.

Kester, Grant. 2008. Collaborative practices in environmental art. In *Artistic Bedfellows: Histories, Theories, and Conversations in Collaborative Art Practices*, 60–63. Lanham, MD: University Press of America.

Kester, Grant. 2011. *The One and the Many: Contemporary Collaborative Art in a Global Context*. Durham, NC: Duke University Press.

Khazaleh, Lorenz. 2013. Growth vs. sustainability: Off to explore one of today's main contradictions. http://www.sv.uio.no/sai/english/research/projects/overheating/news/hylland-eriksen.html (accessed February 17, 2015).

Kim, Sunjung. 2014. On Platform Seoul: Models beyond the biennale. In *Art in the Asia-Pacific: Intimate Publics*, ed. Larissa Hjorth, Natalie King, and Mami Kataoka, 73–89. New York: Routledge.

Kitchin, Rob, Martin Dodge, and Chris Perkins. 2011. Introductory essay: Conceptualising mapping. In *The Map Reader: Theories of Mapping Practice and Cartographic Representation*, ed. Martin Dodge, Rob Kitchin, and Chris Perkins, 1–7. London: Wiley.

Kopomaa, Timo. 2000. *The City in Your Pocket: Birth of the Mobile Information Society*. Helsinki: Gaudemus.

Koskinen, Ilpo. 2007. Managing banality in mobile multimedia. In *The Social Construction and Usage of Communication Technologies: European and Asian Experiences*, ed. Raul Pertierra, 48–60. Singapore: Singapore University Press.

Lammes, Sybille, and Clancy Wilmott. 2013. Mapping the city, playing the city: Location-based apps as navigational interfaces. Paper presented at ICACI (International Cartographic Association) Conference, Dresden, Germany, August, http://icaci.org/icc2013/.

Lapenta, Francesco. 2011. Geomedia: On location-based media, the changing status of collective image production and the emergence of social navigation systems. *Visual Studies* 26 (1): 14–24.

Lasén, Amparo. 2004. Affective technologies: Emotions and mobile phones. *Receiver* 11.

Lee, Dong-Hoo. 2005. Women's creation of camera phone culture. *Fibreculture Journal* 6. http://www.fibreculture.org/journal/issue6/issue6_donghoo_print.html (accessed February 3, 2006).

Lee, Dong-Hoo. 2009. Reimaging urban space: Mobility, connectivity and a sense of place. In *Mobile Technologies*, ed. Gerard Goggin and Larissa Hjorth, 129–135. New York: Routledge.

Lewis, Tania, ed. 2016. *Green Asia: Ecocultures, Sustainable Lifestyles and Ethical Consumption.* London: Routledge.

Lewis, Tania, and Emily Potter, eds. 2011. *Ethical Consumption: A Critical Introduction.* London: Routledge.

Li, Szu-hsien. 2009. Global warming and Chinese myths: A critique of Vincent J. F. Huang's "The Last Penguins and Taiwan Artists in the Global Chinese Context" (*ARTCO Magazine*), March 2009. English translation available at http://www.modernatlantis.com/2011/10/critique.html (accessed February 17, 2015).

Licoppe, Christian. 2004. "Connected" presence: The emergence of a new repertoire for managing social relationships in a changing communication technoscape. *Environment and Planning D* 22 (1): 135–156.

Ling, Richard, and Heather Horst. 2011. Mobile communication in the global South: An introduction. *New Media and Society* 13 (3): 363–374.

Lorimer, Hayden. 2005. Cultural geography: The busyness of being more than representational. *Progress in Human Geography* 29 (1): 83–94.

Maczulak, Anne. 2010. *Conservation: Protecting Our Plant Resources.* New York: Infobase Publishing.

Madianou, Mirca, and Daniel Miller. 2013. Polymedia: Towards a new theory of digital media in interpersonal communication. *International Journal of Cultural Studies* 16 (2): 169–187.

Maharaj, Sarat. 1994. Perfidious fidelity: The untranslatability of the Other. In *Global Visions: Towards a New Internationalism in the Visual Arts*, ed. Jean Fisher, 28–35. London: Kala Press in association with the Institute of International Visual Arts.

Maravillas, Francis. 2006. Cartographies of the future: Asia-Pacific triennials and the curatorial imaginary. In *Eye of the Beholder: Reception, Audience, and Practice of Modern Asian Art*, ed. John Clark, Maurizio Peleggi, and Thiagarajan K. Sabapathy, 244–270. Sydney: Wild Peony.

Martin, Fran, Tania Lewis, and John Sinclair. 2013. Lifestyle media and social transformation in Asia. *Media International Australia* 147:51–61. http://www.findanexpert.unimelb.edu.au/display/publication198665 (accessed February 17, 2015).

Martin, Stewart. 2007. Critique of relational aesthetics. *Third Text* 21 (4): 369–386.

Massey, Doreen. 2005. *For Space*. London: Sage.

Massey, Doreen. 2007. *World City*. Cambridge: Polity Press.

Masters, Coco. 2008. The Japanese way. *Time*, April 17. http://content.time.com/time/specials/2007/article/0,28804,1730759_1734222_1734215,00.html (accessed June 8, 2015).

Maxwell, Richard, and Toby Miller. 2012. *Greening the Media*. Oxford: Oxford University Press.

MCA. 2010. David Stephenson: About the artwork. http://www.mca.com.au/collection/work/201045/ (accessed November 10, 2015).

McQuire, Scott. 2008. *The Media City: Media, Architecture and Urban Space*. London: Sage.

Merewether, Charles. 2007. The freedom of irreverence: Ai Weiwei. *ArtAsiaPacific* (May–June). http://artasiapacific.com/Magazine/53/TheFreedomOfIrreverenceAiWeiwei (accessed February 24, 2014).

Merlen, Geoffrey. 2014. The English biologist Geoffrey Merlen on scientific whaling versus cultural whaling. http://www.seashepherd.org/videos.html (accessed January 20, 2014).

Meskimmon, Marsha. 2011. *Contemporary Art and the Cosmopolitan Imagination*. London: Routledge.

Meskimmon, Marsha. 2013. The precarious ecologies of cosmopolitanism. *Humanities Report* 19 (2): 27–45.

Mirzoeff, Nicholas. 2014. Visualizing the anthropocene. *Public Culture* 26 (2): 213–232.

Mitchell, William J. T. 2005. There are no visual media. *Journal of Visual Culture* 4 (2): 257–266.

Mitsuda, Hisayoshi. 1996. Surging environmentalism in Japan: Back to the root. Presented at the Ninth World Congress of Rural Sociology, Bucharest, July 22–26.

Moores, Shaun. 2012. *Media, Place, and Mobility*. New York: Palgrave Macmillan.

Mørk Petersen, Søren. 2008. Common banality: The affective character of photo sharing, everyday life and produsage cultures. PhD diss., IT University of Copenhagen.

Morley, David. 2005. The domestication of the media and the dis-location of domesticity. In *The Domestication of Media and Technology*, ed. Thomas Berker, Maren Hartmann, Yves Punie, and Katie Ward. New York: McGraw-Hill/Open University Press.

Morton, Timothy. 2007. *Ecology without Nature: Rethinking Environmental Aesthetics*. Cambridge, MA: Harvard University Press.

Morton, Timothy. 2013. *Hyperobjects: Philosophy and Ecology after the End of the World*. Minneapolis: University of Minnesota Press.

Mortreux, Collette, and Jon Barnett. 2009. Climate change, migration, and adaptation in Funafuti, Tuvalu. *Global Environmental Change* 19 (1): 105–112.

Mosquera, Gerardo. 2003. Alien-own/own-alien. In *Complex Entanglements: Art, Globalisation and Cultural Difference*, ed. Nikos Papastergiadis, 18–29. London: Oram Press.

Mullany, Gerry, and Chris Buckley. 2013. Thousands rally in Hong Kong on Tiananmen Square anniversary. *New York Times*, June 4. http://www.nytimes.com/2013/06/05/world/asia/thousands-rally-in-hong-kong-on-tiananmen-square-anniversary.html (accessed February 17, 2015).

Nation. 2015. Mobile Game Asia conference in Bangkok next week. *Nation*, January 13. http://www.nationmultimedia.com/technology/Mobile-Game-Asia-conference-in-Bangkok-next-week-30251801.html (accessed February 17, 2015).

Neri, Corrado. 2013. China has a natural environment, too! Consumerist and ideological eco-imaginaries in the cinema of Feng Xiagang. In *Transnational Ecocinema: Film Culture in an Era of Ecological Transformation*, ed. Tommy Gustafsson and Pietari Kääpä, 119–136. Bristol: Intellect.

Niazi, Tarique. 2013. The Asia-Pacific in the eye of super-storms. *Asia-Pacific Journal* 11 (48) (no. 5, December 2). http://www.japanfocus.org/-Tarique-Niazi/4043 (accessed February 2, 2015).

Nye, Joseph S. 1990. Soft power. *Foreign Policy* 80:153–171.

Otmazgin, Nissim. 2014. *Regionalizing Culture: The Political Economy of Japanese Popular Culture in Asia*. Hawai'i: University of Hawai'i Press.

Palmer, Daniel. 2013. Photography, technology, and ecological criticism: Beyond the sublime image of disaster. In *Environmental Conflict and the Media*, ed. Libby Lester and Brett Hutchins, 75–90. New York: Peter Lang.

Palmer, Daniel. 2014. Overflow: Tales of the sublime. In *The Nineteenth Biennale of Sydney: You Imagine What You Desire*, ed. Juliana Engberg. Sydney.

Papastergiadis, Nikos. 2011. Collaboration in art and society: A global pursuit of democratic dialogue. In *Globalization and Contemporary Art*, ed. Harris J. West, 275–288. Sussex: Wiley-Blackwell.

Papastergiadis, Nikos. 2012. *Cosmopolitanism and Culture*. Cambridge: Polity Press.

Parikka, Jussi. 2012. Introduction: The materiality of media and waste. In *Medianatures: The Materiality of Information Technology and Electronic Waste*, ed. Jussi Parikka. Ann Arbor: Open Humanities Press. http://www.livingbooksaboutlife.org/books/Electronic_waste/Introduction (accessed February 17, 2015).

PBS Frontline. 2011. The atomic artists: "Art cannot be powerless," an interview with Chim↑Pom's leader Ryuta Ushiro. *PBS Frontline*, July 26. http://www.pbs.org/wgbh/pages/frontline/the-atomic -artists/art-cannot-be-powerless (accessed January 15, 2014).

Pertierra, Raul. 2000. The market in Asian values. In *Asian Values: Encounter with Diversity*, ed. Josiane Cauquelin, Paul Lim, and Birgit Mayer-König, 118–138. New York: Routledge.

Pink, Sarah. 2011. Sensory digital photography: Re-thinking "moving" and the image. *Visual Studies* 26 (1): 4–13.

Pink, Sarah. 2012. *Situating Everyday Life: Practices and Places*. London: Sage.

Pink, Sarah, and Larissa Hjorth. 2012. Emplaced cartographies: Reconceptualising camera phone practices in an age of locative media. *Media International Australia* 145 (November): 145–155.

Pink, Sarah, and Tania Lewis. 2014. Making resilience: Everyday affect and global affiliation in Australian slow cities. *Cultural Geographies*, February. http://cgj.sagepub.com/content/early/2014/ 02/04/1474474014520761

Pink, Sarah, and Kirsten Leder Mackley. 2013. Saturated and situated: Rethinking media in everyday life. *Media Culture and Society* 35 (6): 677–691.

Postill, John. 2011. *Localizing the Internet*. Oxford: Berghahn.

Postill, John. 2014. A critical history of Internet activism and social protest in Malaysia, 1998– 2011. *Asiascape: Digital Asia Journal* 1–2:78–103. http://rmit.academia.edu/JohnPostill/Papers (accessed February 17, 2015).

Qiu, Jack L. 2012. Network labor: Beyond the shadow of Foxconn. In *Studying Mobile Media*, ed. Larissa Hjorth, Jean Burgess, and Ingrid Richardson, 173–189. New York: Routledge.

Radio Netherlands Worldwide (RNW). 2013. Ducks not tanks, says Yu, creator of Weibo, banned June 4th meme. http://www.rnw.org/archive/ducks-not-tanks-says-yu-creator-weibos-banned -june-4th-meme.

Rafael, Vincent. 2003. The cell phone and the crowd: Messianic politics in the contemporary Philippines. *Popular Culture* 15 (3): 399–425.

Rancière, Jacques. 2009. *Aesthetics and Its Discontents*. Trans. S. Corcoran. Cambridge: Polity.

Raunig, Gerald. 2007. *Art and Revolution: Transversal Activism in the Long Twentieth Century*. Trans. Aileen Derieg. Los Angeles: Semiotext(e).

Robison, Richard, and David S. G. Goodman. 1996. *The New Rich in Asia*. London: Routledge.

Schneider, Arnd, and Christopher Wright. 2013. *Anthropology and Art Practice*. Oxford: Bloomsbury.

Sea Shepherd. 2014. http://seashepherd.unitee.com.au/about-sea-shepherd.

Seno, Alexandra. 2009. Art wars: Hong Kong vs. Singapore. *Wall Street Journal*, October 30. http://online.wsj.com/news/articles/SB125678376301415081 (accessed February 5, 2014).

Sharp, Kristen. 2010. Travelling the distance. In *Outer Site: The Intercultural Projects of RMIT Art in Public Space*, ed. Geoff Hogg and Kristen Sharp, 26–39. Balnarring: McCulloch & McCulloch.

Sheller, Mimi. 2004. Mobile publics: Beyond the networked perspective. *Environment and Planning D: Society and Space* 22:39–52.

Sheller, Mimi. 2014. Mobile art: Out of your pocket. In *The Routledge Companion to Mobile Media*, ed. Gerard Goggin and Larissa Hjorth, 197–205. New York: Routledge.

Shiva, Vandana. 2002. *Water Wars: Privatization, Pollution and Profit*. London: Pluto Press.

Shizimu, Hiroko. 2012. A new perspective: Ichi Ikeda. *Women Environmental Artists Directory*. http://weadartists.org/japanese-anti-nuclear-art-ichi-ikeda-interview (accessed February 4, 2015).

Sicart, Miguel. 2014. *Play Matters*. Cambridge, MA: MIT Press.

Silverstone, Roger, and Eric Hirsch, eds. 1992. *Consuming Technologies: Media and Information in Domestic Spaces*. London: Routledge.

Singapore Biennale. 2013. Curators. http://www.singaporebiennale.org/curators.html (accessed February 17, 2015).

Slater, David H. 2011. Fukushima women against nuclear power: Finding a voice from Tohoku. *Asia-Pacific Journal: Japan Focus*. http://japanfocus.org/events/view/117 (accessed February 4, 2015).

Slater, David H., Nishimura Keiko, and Love Kindstrand. 2012. Social media, information and political activism in Japan's 3.11 crisis. *Japan Focus*. http://japanfocus.org/-Nishimura-Keiko/3762.

Smart, Carol. 2007. *Personal Life*. Oxford: Polity Press.

Smith, Bridie. 2011. Fates conspire to concoct a recipe for disaster. *Age*. http://www.theage.com.au/environment/fates-conspire-to-concoct-a-recipe-for-disaster-20110111-19mp7.html (accessed January 8, 2014).

Smith, Terry. 2011a. *Contemporary Art World Currents*. London: Laurence King.

Smith, Terry. 2011b. Currents of world-making in contemporary art. *World Art* 2:20–36.

Stimson, Blake, and Gregory Sholette. 2007. *Collectivism after Modernism: The Art of Social Imagination after 1945*. Minneapolis: University of Minnesota Press.

Strengers, Yolanda, Cecily Maller, Larissa Nicholls, and Sarah Pink. 2014. Australia's rising air-con use makes us hot and bothered. *Conversation* 2 (January). http://theconversation.com/australias-rising-air-con-use-makes-us-hot-and-bothered-20258 (accessed October 2, 2015).

Tacchi, Jo, Heather Horst, Evangelia Papoutsaki, Verena Thomas, and Joys Eggins. 2013. *PACMAS State of Media and Communication Report 2013*. Melbourne: ABC International Development.

http://www.pacmas.org/profile/pacmas-state-of-media-and-communication-report-2013 (accessed February 17, 2015).

Takahashi, Dean. 2014. Only 0.15 percent of mobile gamers account for 50 percent of all in-game revenue. *Venture Beat*. http://venturebeat.com/2014/02/26/only-0-15-of-mobile-gamers-account-for-50-percent-of-all-in-game-revenue-exclusive (accessed February 17, 2015).

Tan, Lawrence. 2008. The Himalayas Center in Pudong, Shanghai. *Luxury Insider*. http://www.luxury-insider.com/luxury-news/2008/11/himalayas-center-in-pudong-shanghai (accessed February 24, 2014).

Taylor, Jonathan. 1999. Japan's global environmentalism: Rhetoric and reality. *Political Geography* 18 (5): 535–562.

Taylor, Nora. 2011. Art without history? Southeast Asian artists and their communities in the face of geography. *Art Journal* 70 (2): 7–23.

Taylor, Nora. 2012. *Modern and Contemporary Southeast Asian Art: An Anthology.* Ithaca, NY: Cornell University Press.

Teck-Peng, Chang, and Hu Yong. 2011. *Social Media Uprising in the Chinese-Speaking World.* Hong Kong: In-Media.

Thrift, Nigel. 2008. *Non-representational Theory: Space, Politics, Affect.* London: Routledge.

Turner, Bryan. 2007. The enclave society: Towards a sociology of immobility. *European Journal of Social Theory* 10 (2): 287–304.

Uricchio, William. 2011. The algorithmic turn: Photosynth, augmented reality, and the changing implications of the image. *Visual Studies* 26 (1): 25–35.

Venkatraman, Radha. 2012. A curatorial perspective on Roundtable: The Ninth Gwangju Biennale. *Asia Society*, October 20. http://asiasociety.org/india/curatorial-perspective-roundtable-9th-gwangju-biennale (accessed February 26, 2014).

Verhoeff, Nanna. 2013. The medium is the method: Locative media for digital archives. In *(Dis)Orienting Media and Narrative Mazes*, ed. Julia Eckel, Bernd Leiendecker, Daniela Olek, and Christine Piepiorka, 17–30. Bielefeld: Transcript.

Wajcman, Judy, Michael Bittman, and Jude Brown. 2009. Intimate connections: The impact of the mobile phone on work/life boundaries. In *Mobile Technologies*, ed. Gerard Goggin and Larissa Hjorth, 9–22. London: Routledge.

Warner, Michael. 2002. Publics and counterpublics. *Public Culture* 14 (1): 49–90.

WAWA Project. 2014. http://wawa.or.jp/en.

Weiner, Jonah. 2013. The artist who talks with the fishes. *New York Times*, June 28. http://www.nytimes.com/2013/06/30/magazine/the-artist-who-talks-with-the-fishes.html?pagewanted=all&_r=0 (accessed February 4, 2015).

Wells, Lyn. 2003. *The Photography Reader*. London: Sage.

Wheeler, Wendy. 2011. The biosemiotic turn: Abduction, or The nature of creative reason in nature and culture. In *Ecocritical Theory: New European Approaches*, ed. Axel Goodbody and Kate Rigby, 270–282. Charlottesville: University of Virginia Press.

Wilken, Rowan. 2005. From stabilitas loci to mobilitas loci: Networked mobility and the transformation of place. *Fibreculture Journal* 6:1–16.

Wilken, Rowan, and Gerard Goggin. 2012. *Mobile Technology and Place*. New York: Routledge.

Williams, Linda. 2013. Affective poetics and public access: The critical challenges of environmental art. *Australasian Journal of Ecocriticism and Cultural Ecology* 3:16–30.

Williams, Linda. 2014. Reconfiguring place: Art and the global imaginary. In *The Sage Handbook of Globalization*, ed. Manfred Steger, Paul Battersby, and Joseph Siracusa, 463–480. London: Sage.

Wilmott, Clancy. 2012. Cartographic city: Mobile mapping as a contemporary urban practice. *Refractory: A Journal of Entertainment Media* 21 (December 28). http://refractory.unimelb.edu .au/2012/12/28/wilmott.

Wilson, Rob. 2000. Imagining "Asia-Pacific": Forgetting politics and colonialism in the magical waters of the Pacific: An Americanist critique. *Cultural Studies* 14 (3–4): 562–592.

Wirsing, Robert, Daniel Stoll, and Christopher Jasparro. 2013. *International Conflict over Water Resources in Himalayan Asia*. Critical Studies of the Asia Pacific Series. New York: Palgrave Macmillan.

Wong, Tai-Chee, and Belinda Yuen. 2011. *Eco-city Planning: Policies, Practice and Design*. New York: Springer.

Xiaodong, Liu, and Phillip Tinari. 2012. Truth in painting does not equal truth in reality. *Parkett Art* 91. http://www.parkettart.com/downloadable/download/sample/sample_id/327 (accessed February 4, 2015).

Yao, Pauline J. 2012. Dark matter. *Artforum*, January. http://www.lehmannmaupin.com/artists/ liu-wei/press/1520/artist_video (accessed February 4, 2015).

Yoshinko, Kosaku. 1992. *Cultural Nationalism in Contemporary Japan: A Sociological Enquiry*. London: Routledge.

Zeplin, Pamela. 2013. "The liquid continent": Globalization, urbanization, contemporary Pacific art, and Australia. In *Re-imagining the City: Art, Globalization, and Urban Spaces*, ed. Elizabeth Grierson and Kristen Sharp, 133–152. Bristol: Intellect.

Zylinska, Joanna. 2014. *Minimal Ethics for the Anthropocene*. Ann Arbor: Open Humanities Press.

Index

VIP*

who work with

FARM and EARTH-MOVING MACHINES

by

Freeman, Westover, and Willis

Illustrated by Harry Garo

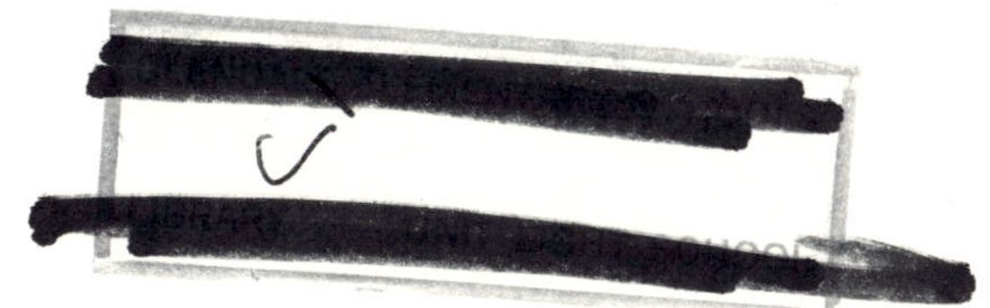

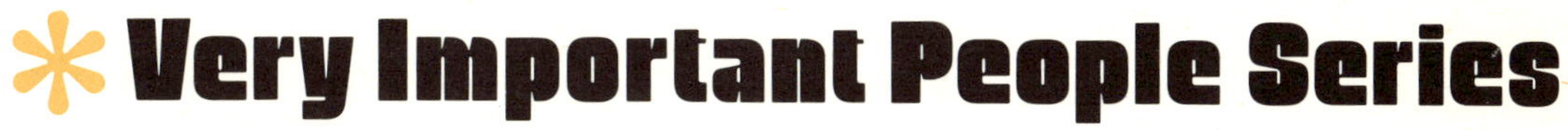

 **Very Important People Series**

AN ELK GROVE BOOK

CHILDRENS PRESS, CHICAGO

Library of Congress Cataloging in Publication Data

Freeman, Dorothy Rhodes.
   Very important people who work with farm and
earth-moving machines.

   (VIP series, set I, book 2)
   SUMMARY: Describes the jobs and training of the
people who operate and service various types of farm
and earth-moving machinery.

   "An Elk Grove book."

   1. Earthmoving machinery operators—Vocational
guidance—Juvenile literature.   2. Agricultural
machinery operators—Vocational guidance—Juvenile
literature.   [1. Earth-moving machinery operators—
Vocational guidance.   2. Agricultural machinery
operators—Vocational guidance]   I. Westover,
Margaret, joint author.   II. Willis, Willma, joint
author.   III. Garo, Harry, illus.   IV. Title.
V. Farm and earth-moving machines.
TA725.F68   624'.152'023   72-10352
ISBN 0-516-07453-9

2 3 4 5 6 7 8 9 10 11 12 13 14 15 16 17 18 19 20 21 22 23 24 25 R 75 74

VIP*

who work with

FARM and EARTH-MOVING MACHINES

SET 1 – BOOK 2

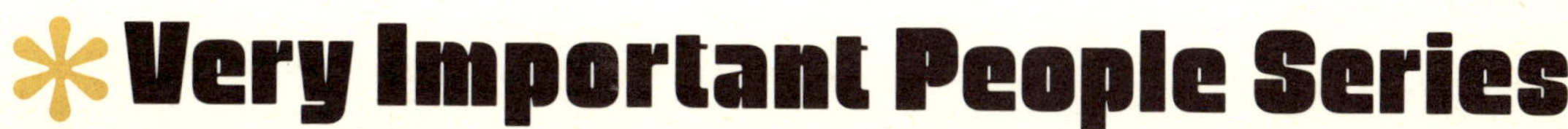

V.I.P. SERIES

CONTENTS

If you work with FARM or EARTH-MOVING MACHINES, you could be . . .

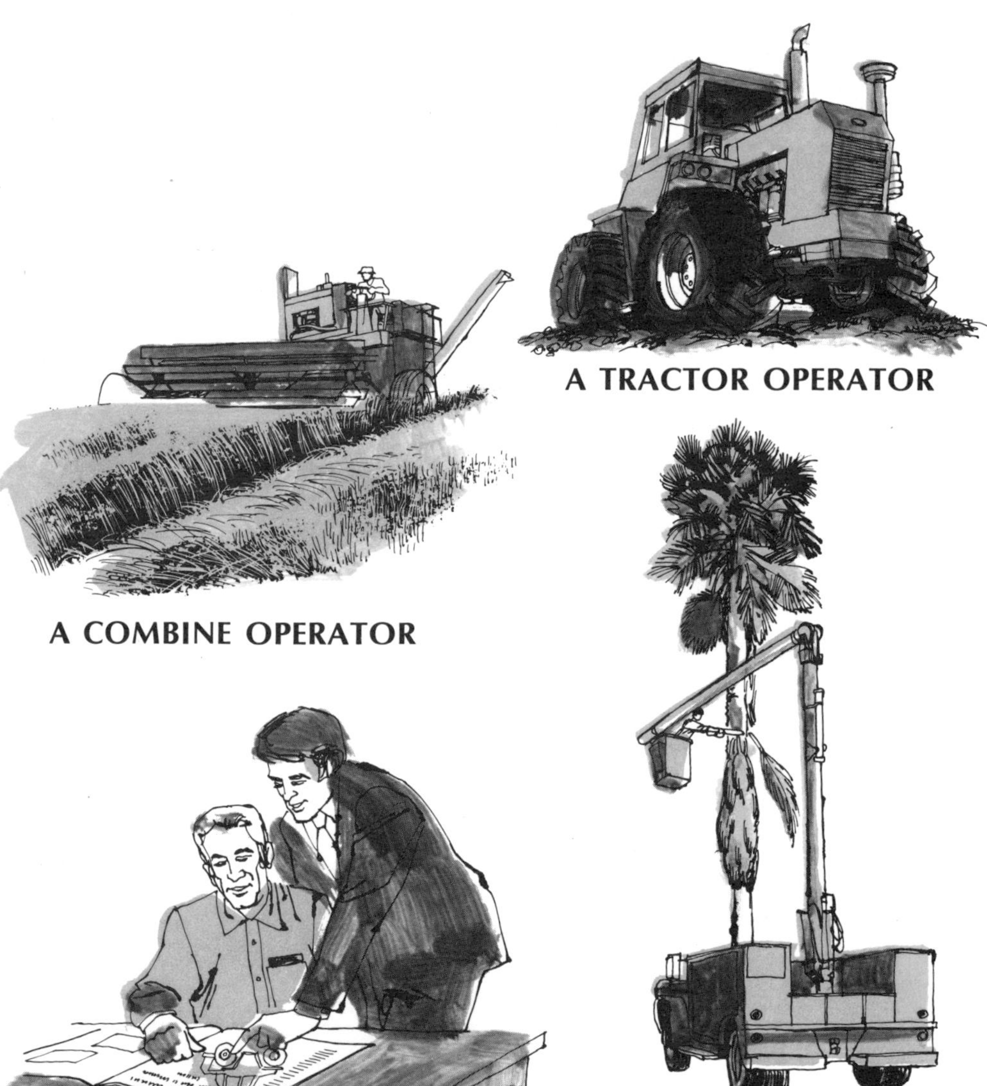

A COMBINE OPERATOR

A TRACTOR OPERATOR

A FARM MACHINERY SALESMAN

A TREE TRIMMER OPERATOR

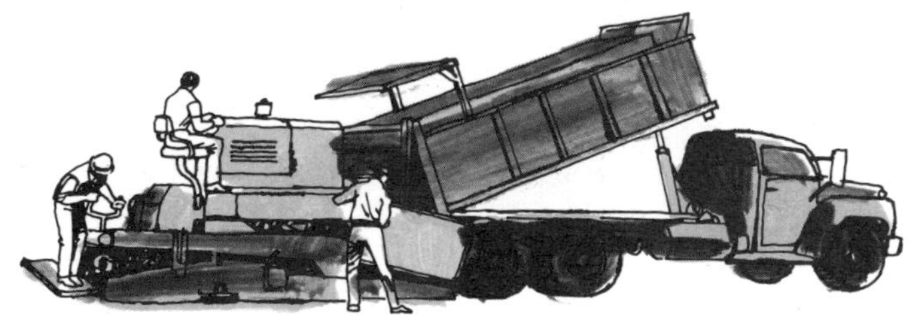

**A ROAD PAVER
OPERATOR**

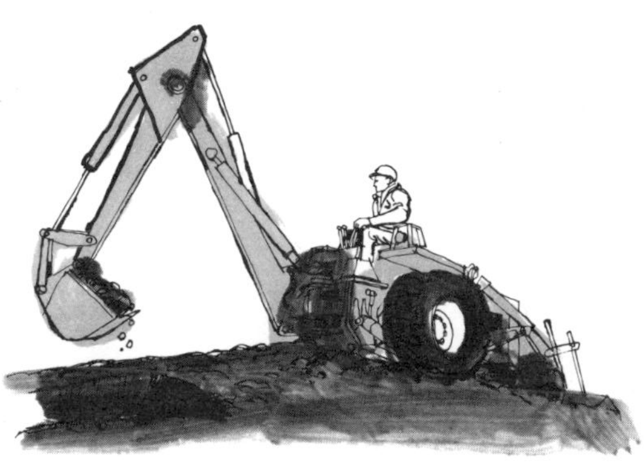

**A LOADER-BACKHOE
OPERATOR**

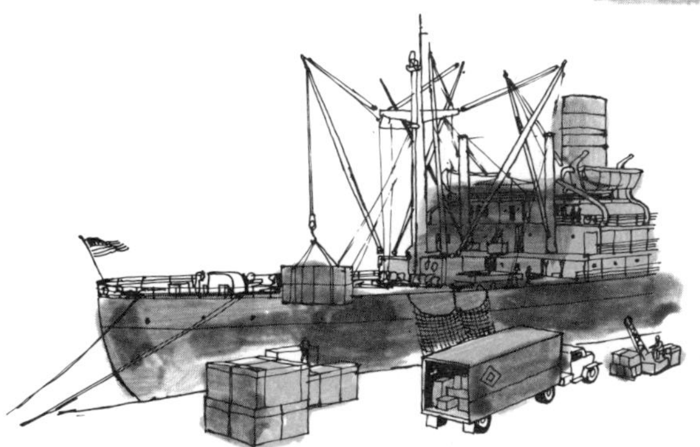

A DERRICK OPERATOR

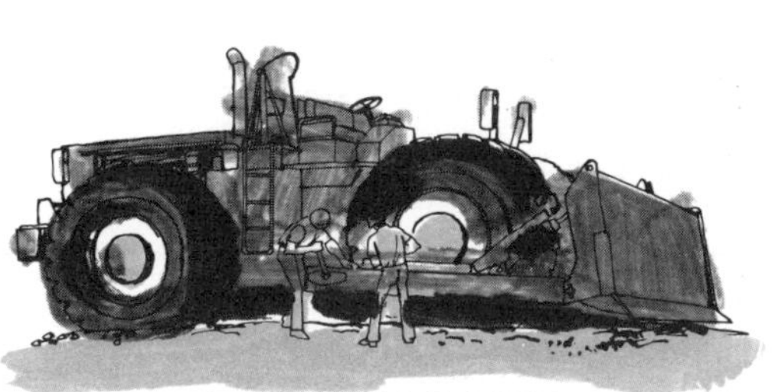

**A HEAVY DUTY
EQUIPMENT REPAIRMAN**

CHOOSING YOUR WORK

There are so many exciting jobs in working with machines that it may be hard for you to choose which jobs you'd like the most.

Here is a way to help you think about it. Read through these questions and answer them for yourself:

1. Do you like to work with your hands? (Do you like to make models? Draw pictures? Do puzzles?)
2. Do you like to work with numbers and lists? (Do you like to gather facts about baseball players? Memorize the states in the U.S.A.? Do you like to use maps?)
3. Do you like to work with people? (Do you like to be on committees? Sell things for the Scouts? Meet new people?)

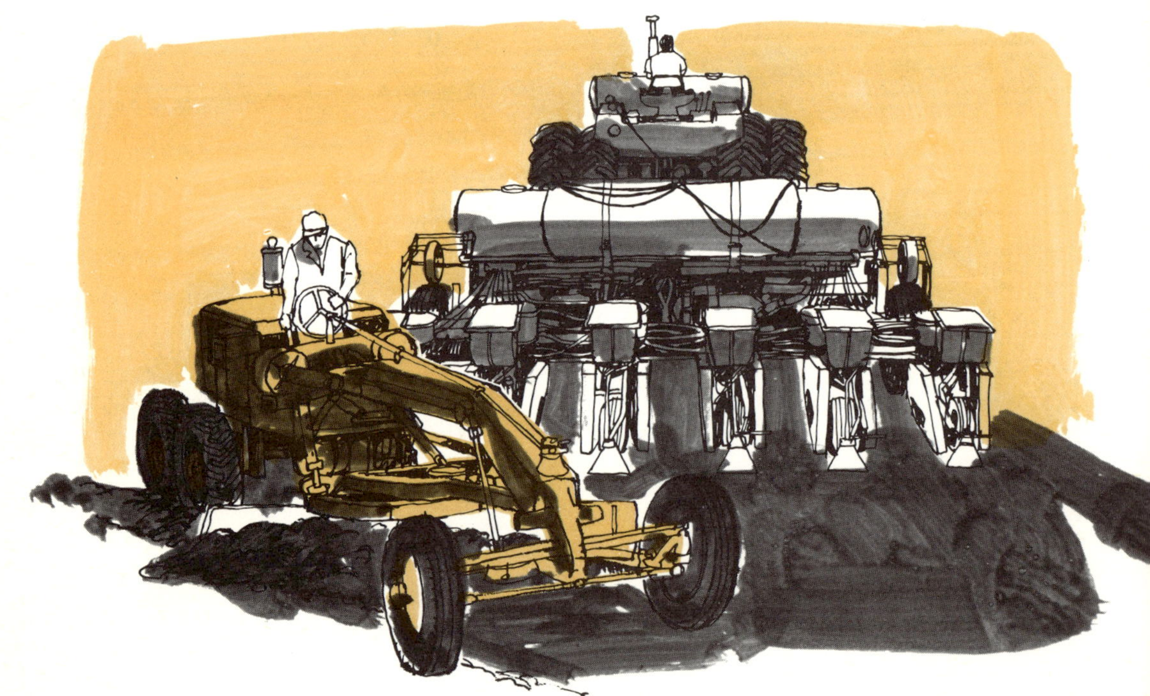

Which of the three groups of activities do you like most? Your answers give you clues to your interests.

Were your interests more in

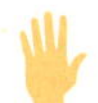 1. Hands?

 2. Numbers and lists?

 3. People?

Most of the jobs in this book are for people who like to work with their hands. They drive farm and earth-moving machines. The machines may be as simple as a small *tractor* or as complicated as a huge *crane*. Look for the hand symbol by these jobs in the rest of this book.

There are also jobs for people who like to work with numbers and lists. A *bulldozer operator* has to be able to read numbers that tell him how much earth to move. Numbers and lists give information called *data*. Look for this symbol to find these jobs.

Workers who like to deal with people might get jobs as supervisors of other workers, or they could become equipment salesmen. A people symbol will show these jobs.

Of course, many jobs require more than one interest. The symbols tell you the most important interests for each job. When you see more than one symbol, the first is more important than any that follow.

This book tells you about the V.I.P. who work with farm and earth-moving machines.

V—Very
I—Important
P—People

V.I.P. WHO WORK WITH FARM MACHINES

There are farms in every state in the United States, including Alaska and Hawaii. Using modern machines, these farms produce hundreds of different foods and other important crops.

On the smallest farms, the farmer and his family run all the equipment. On large farms, many workers are needed to operate and repair the machines.

Farm equipment operators are V.I.P. (Very Important People). Using tractors and many other machines, they grow seven times as much food as farmers did when your grandfather was young.

The size of the farm has a lot to do with the kind of work an equipment operator does. On a very large farm, equipment operators can keep busy all year.

If the farm or ranch is in a warm climate, workers can plant and harvest crops year round. In a cold climate they help repair equipment or buildings in the winter time.

But on a smaller farm, there may not be enough work to keep an equipment operator busy all year. He may have to leave the farm for part of the year and take another job such as truck driving or factory work.

Farm equipment operators find life on small and large farms different in other ways. A worker on a small farm usually does not live on the farm, but he goes there to work during the day.

A worker on a large farm usually lives in a house or a *mobile home* right on the farm. Large farms often have swimming pools, hunting grounds and picnic areas for employees and their families.

On a small farm an operator works early and late, for as many hours as he is needed. On a large farm an operator is more likely to have regular hours and get vacations with pay.

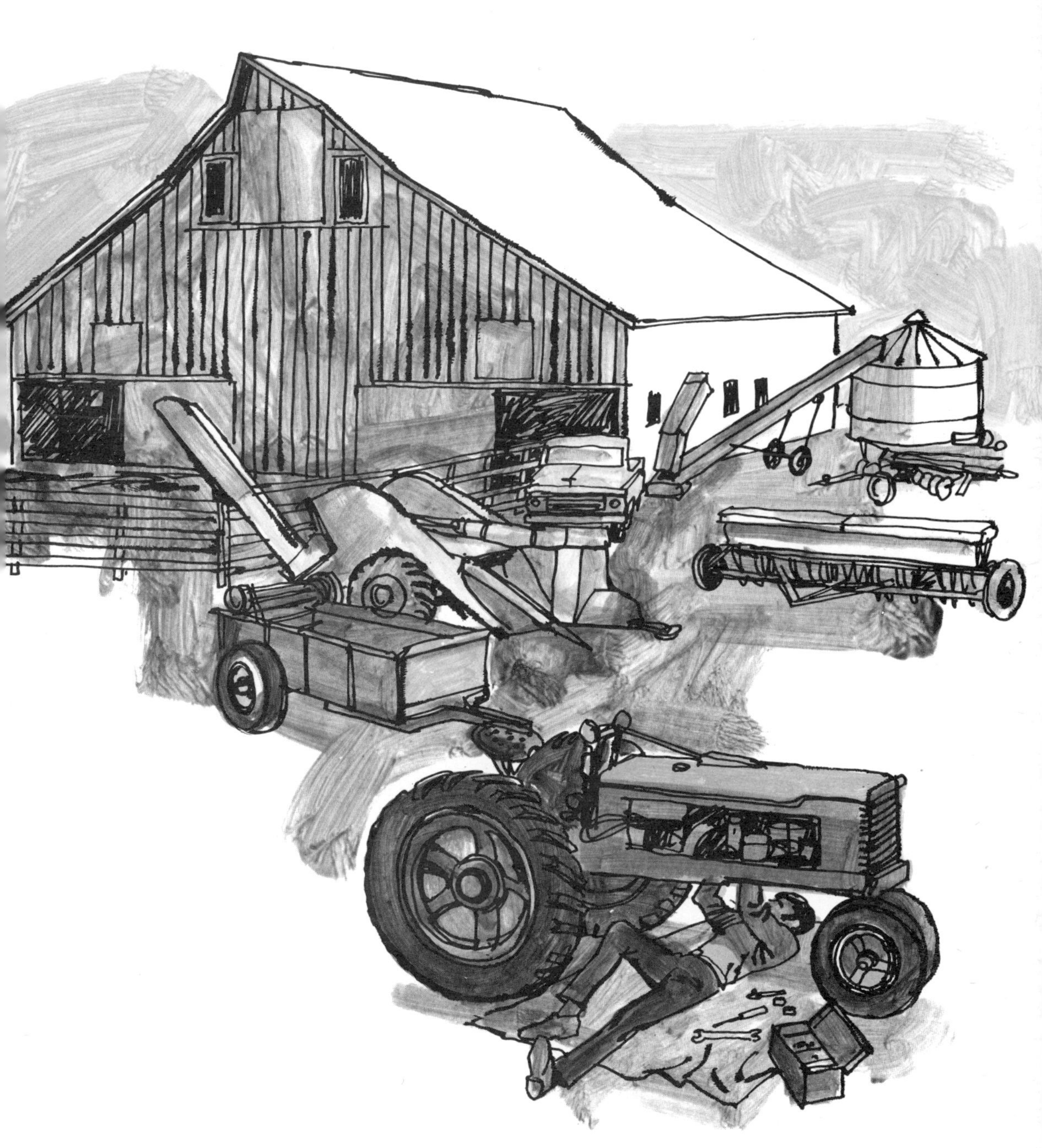

TRACTOR OPERATORS

The tractor is the most important of all farm machines.

There are two basic kinds of tractors. One has rubber-tired wheels. The other has continuous metal *tread* like an army tank.

Many tractors have *four-wheel drive*. With power to each wheel, the driver can get the machine out of ruts. He can work on wet or hilly ground without slipping.

Some farm tractors are deluxe. Their air-conditioned cabs are enclosed with glass. The operator can work in cold, windy, or very hot weather. He is protected from dust, sprays, and insects.

The largest tractors have many *gages* and *dials*. For instance, a gage called a *pyrometer* shows the temperature of the exhaust. A dial tells the operator the angle of the rear wheels, so he can guide the tractor safely on steep hills.

Of course, a tractor can't do the farm work by itself. A skilled operator uses it to pull different *attachments* to do most of the farming jobs.

With a tractor and its attachments, one man can do more work than several men used to do with animals and hand tools.

The next few pages tell more about farm jobs and the tractor attachments that do them.

This information may help you decide whether you would like to be a farm equipment operator.

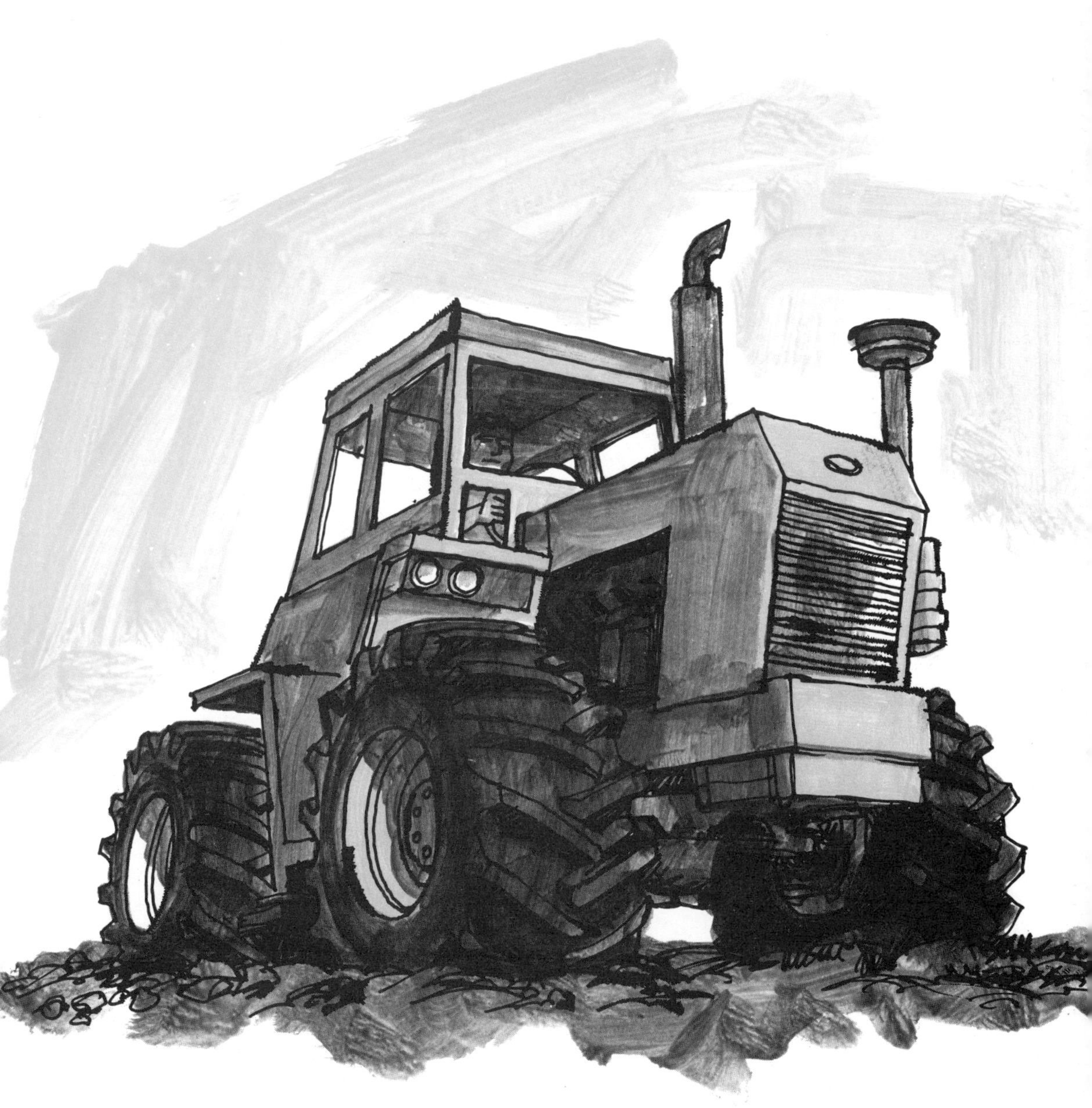

PREPARING THE SOIL

To prepare the earth for planting, a tractor operator sometimes pulls a *spring tooth harrow.* This attachment digs deep and tears up old roots.

With an attachment called a *disk harrow,* the tractor operator cuts, turns, and breaks the soil. This gets the soil ready for planting. Some disk harrows can prepare a strip 33 feet wide at one time.

The tractor operator can tow a *plow* that cuts a groove, called a *furrow,* in the earth. Furrows catch and hold water for the growing plants.

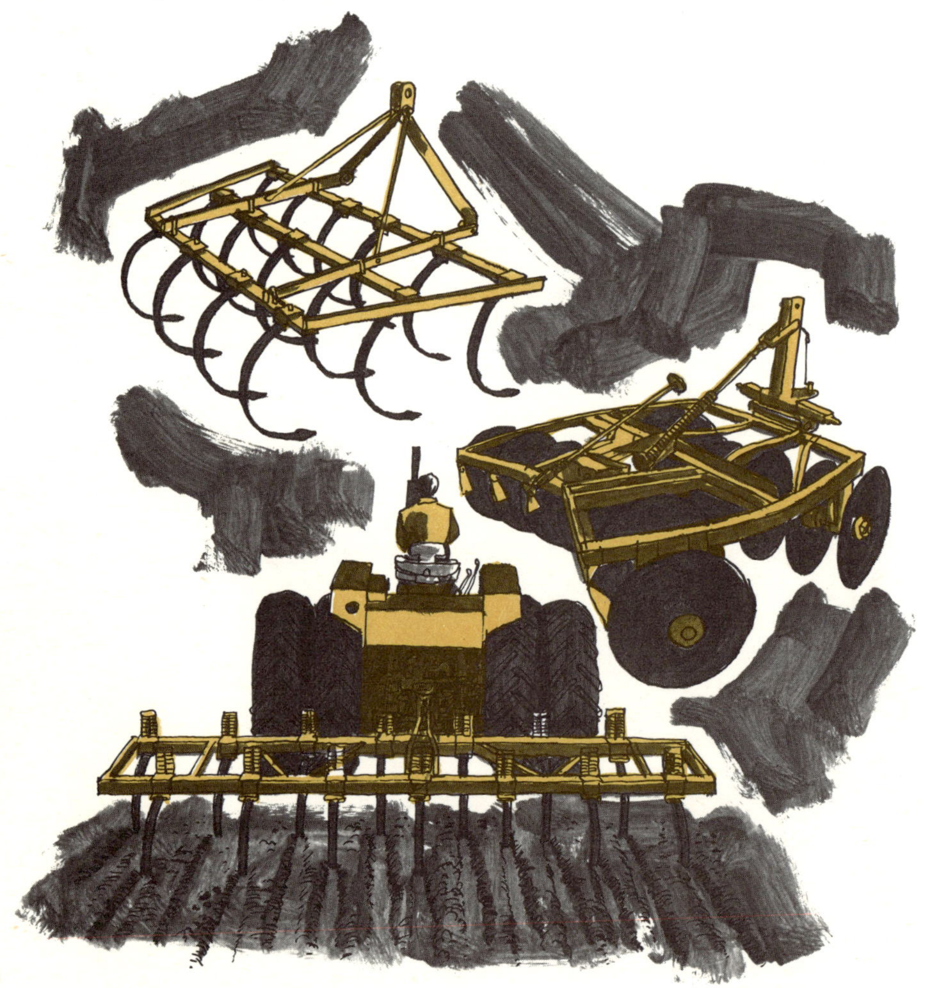

PLANTING THE CROPS

Dragging a planting attachment, the tractor driver can put seeds or young plants into the soil. Young plants are called *starts*.

To plant celery, a helper rides on the planting attachment and puts the starts into little pockets in a big wheel. As the tractor pulls the planter, the wheel turns and puts the celery plants into the earth.

Some large attachments do several jobs at once. One machine can put *fertilizer, weed killer,* and *insecticide* into the soil at the same time that it plants the seeds. This machine can plant 12 rows of corn at a time.

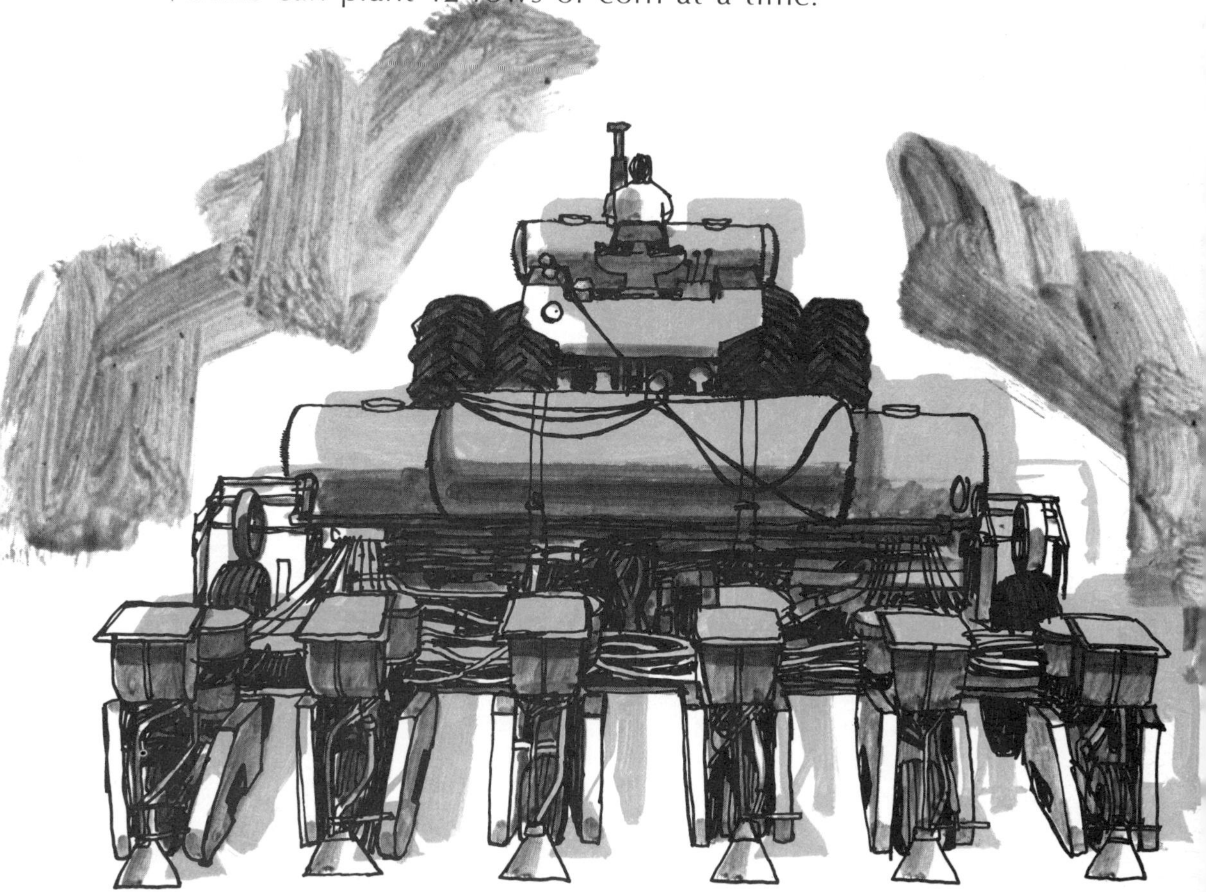

CULTIVATING

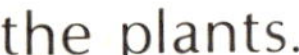

While the crops are growing, the tractor operator pulls attachments through the rows of plants to chop down weeds or loosen the soil. This is called *cultivating*. After this is done, water and air can get down into the soil.

The growing crops are food for insects as well as for people. To prevent insects from eating the crops, the tractor driver uses a power-driven pump to spray insecticides on the plants.

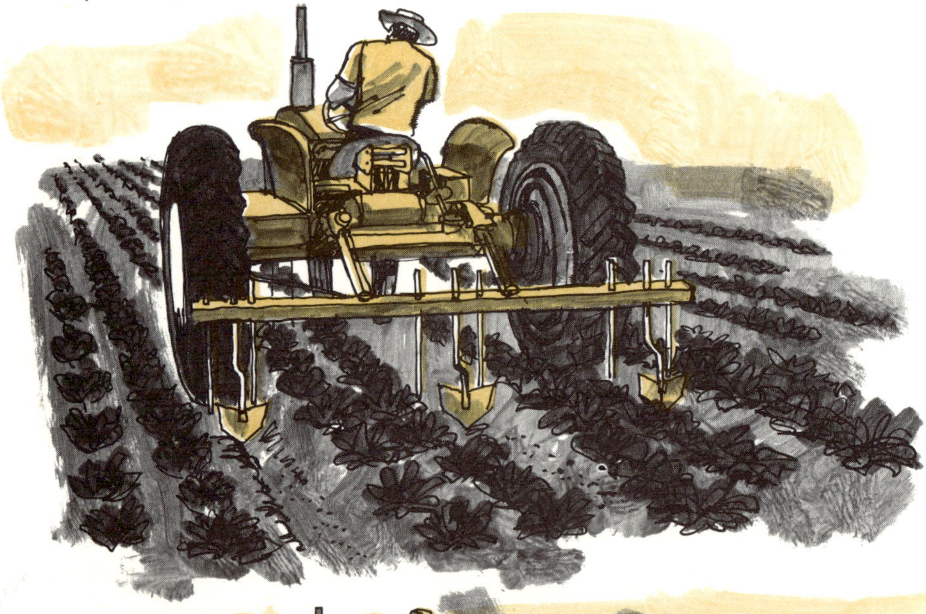

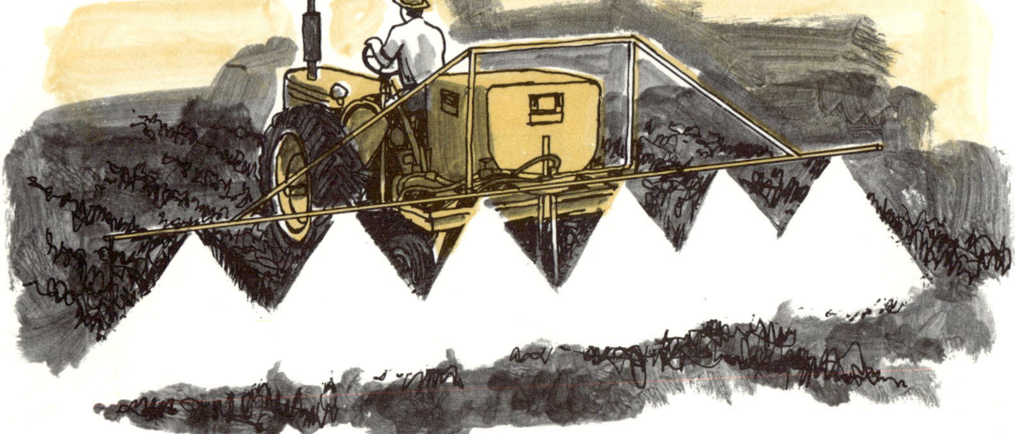

HARVESTING MACHINE OPERATORS

Picking or gathering crops from the fields is called harvesting.

A *combine harvester* does three jobs at once. It cuts the wheat, removes the grain from the stalks, and loads the grain into a truck.

Some **COMBINE OPERATORS** specialize in working with these machines. The operators and helpers are called harvesting crews. They travel from the Mexican border north to Canada, harvesting wheat or other grains in field after field.

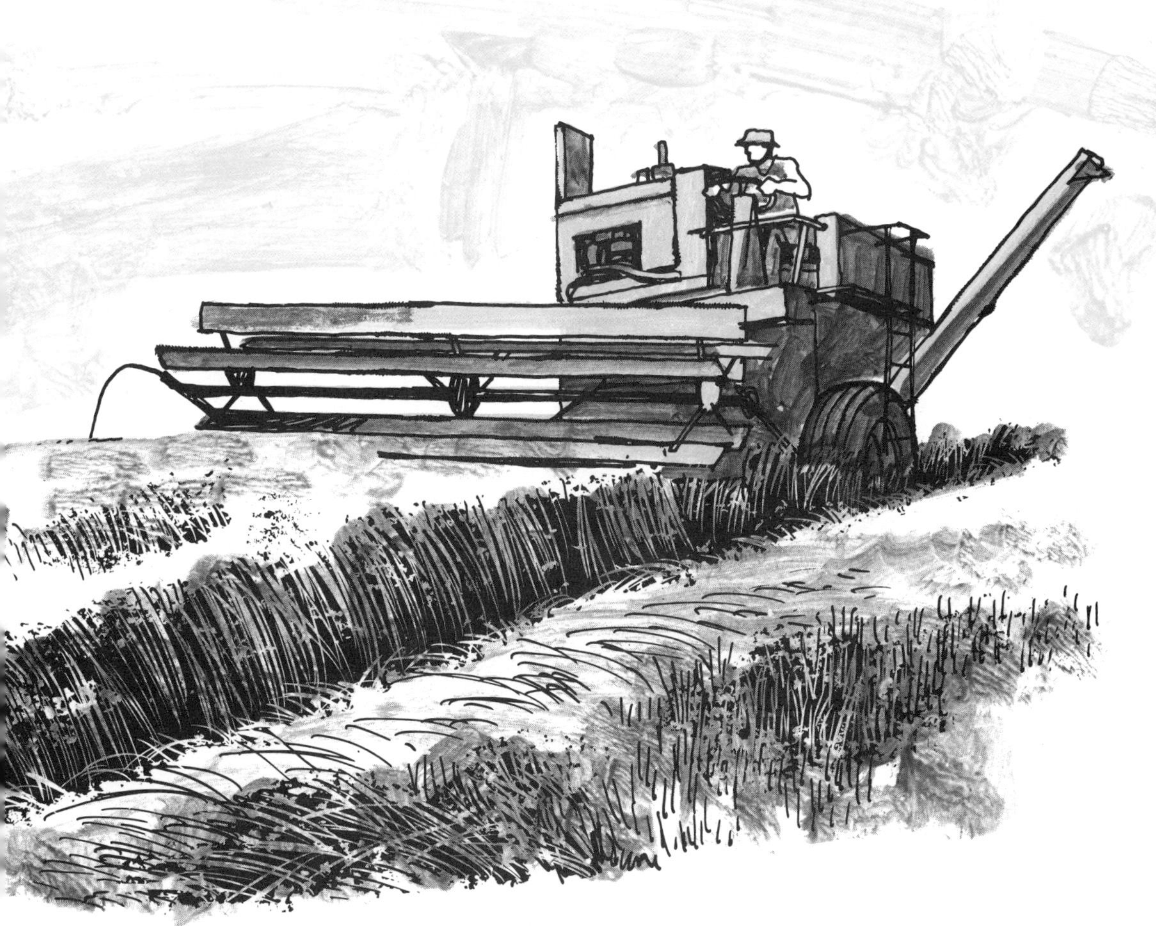

Some fruits are also harvested mechanically. An operator holds a machine that grabs the tree trunk and shakes it, making the fruit fall into a catching screen. Cherries and prune plums are harvested this way.

The fruit travels on a *conveyor belt* from the catching screen to boxes for loading.

When potatoes are mechanically harvested, the machine cuts off the stems and puts the potatoes into a bin. At the same time, the machine grinds up the stems and roots of the plant. These are then plowed back into the soil to help the next crop grow.

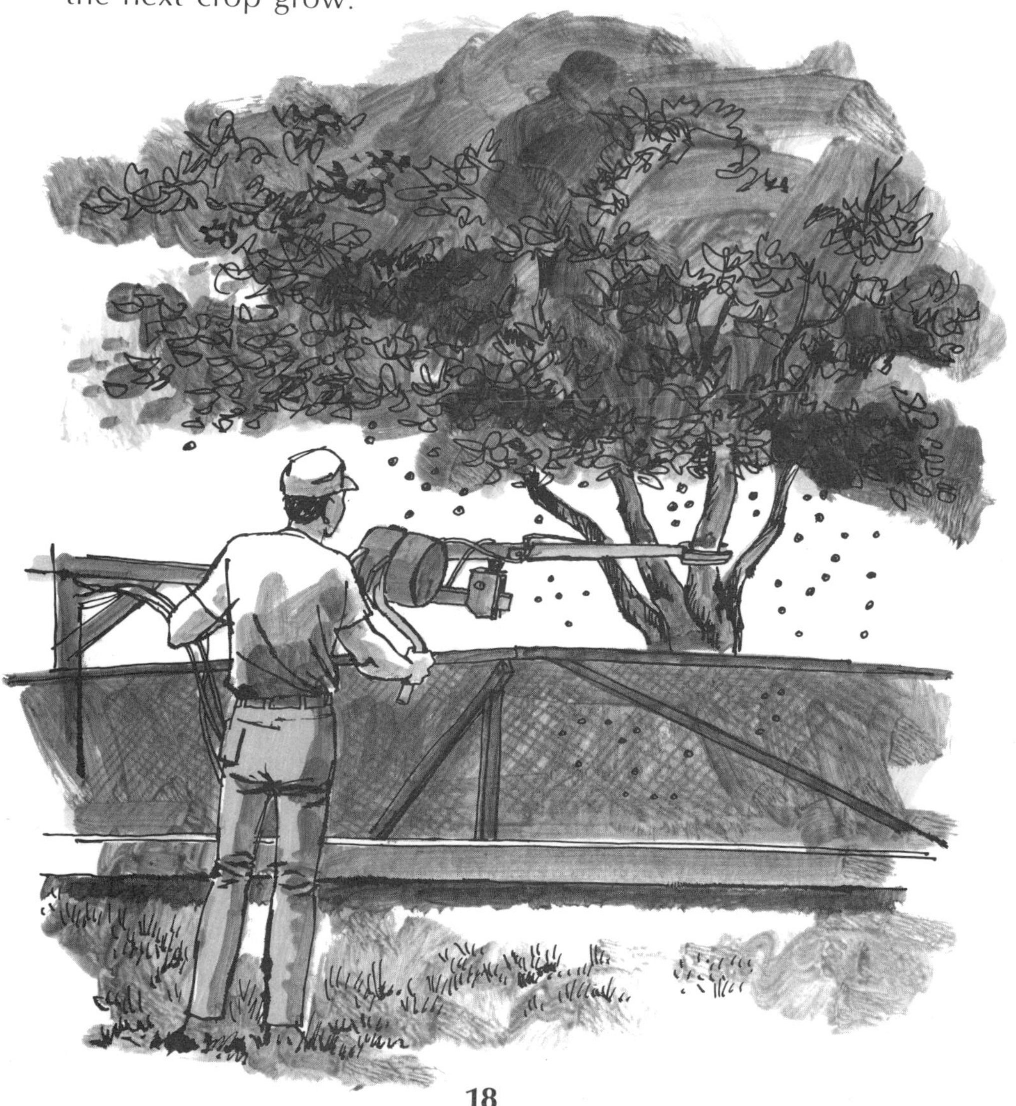

LEARNING THE JOB AND GETTING IT

Most people learn how to operate tractors and farm machinery from friends who are farmers. A few high schools and community colleges teach classes in farm equipment operation.

Most farmers prefer to hire workers they know or ones who have lived on a farm and like country life. But if farm labor is scarce, a beginner may find a job even if he doesn't know the farmer.

Young people may find it easier to get jobs on farms if they have belonged to *4-H* or *Future Farmers of America*.

A farmer likes to hire a person who

- likes to live in the country
- is dependable and can plan his work well
- has skills in working with machines
- is over 18 years old (unless he is well known to the farmer and very capable)
- may have taken agriculture classes in high school or junior college
- is in good physical condition, with a strong back and no allergies
- has good eyesight and good coordination.

FARM MECHANICS

Some mechanics specialize in farm machinery repair. They take care of the gasoline and *diesel* engines, and repair the tractor and its attachments. They also repair other machinery such as milking machines.

Every farm has some kind of repair shop. This shop may be part of a barn where the tractors and attachments are stored. Or it may be a large, modern workshop.

The farmer who has his own **FARM MECHANIC** can usually have his machinery worked on without delay. This is very important because crops may be lost if the equipment isn't ready to use at the right time.

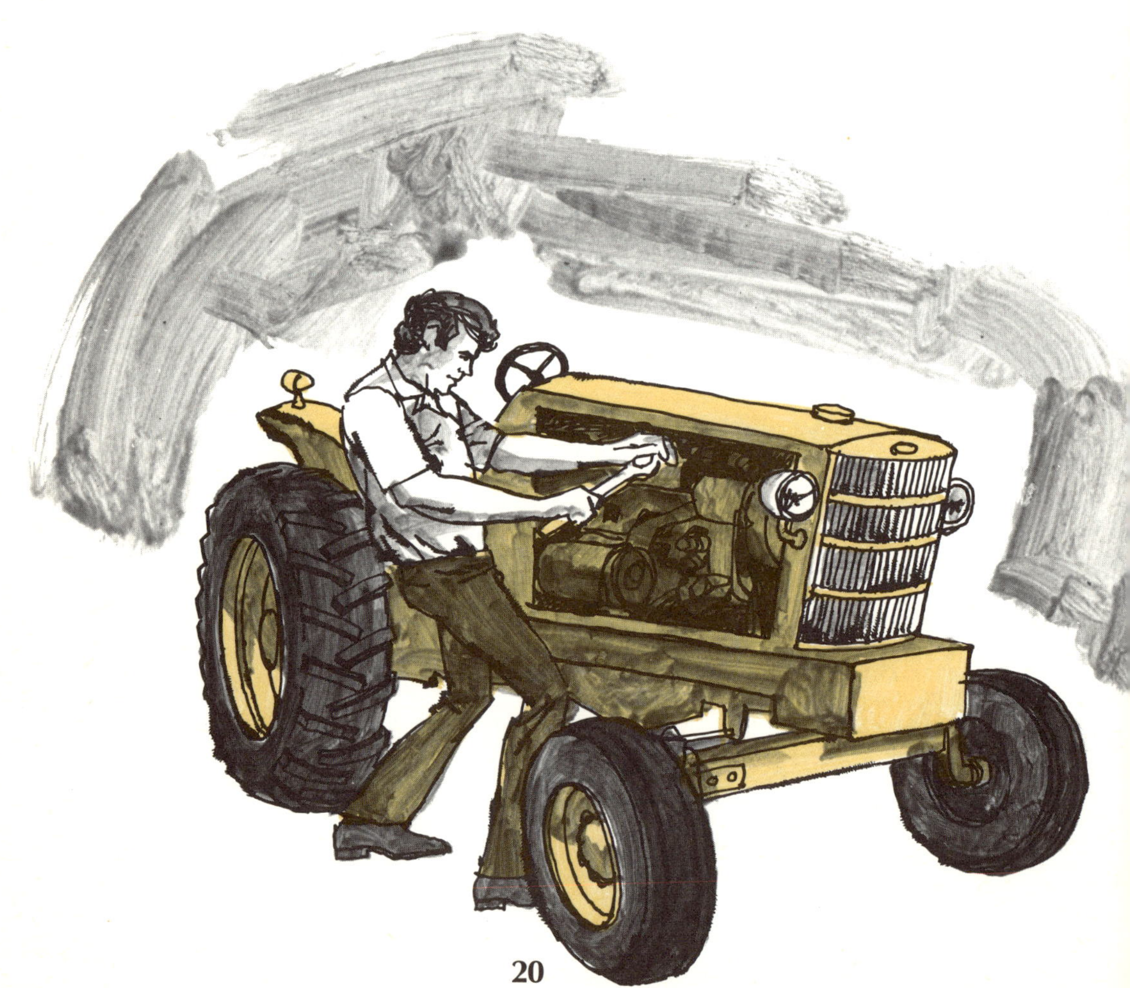

Very large farms employ several mechanics. Each mechanic may specialize in one kind of job.

Large farms sometimes have fields that are many miles apart. A mechanic for such a farm might have to travel with his equipment to repair a machine that is far away.

When he is far from the farm repair shop, the mechanic may have to use his ordinary tools in new and different ways to get the job done.

Farmers who don't hire their own mechanics can call a mechanic to come to the farm. They may also take the broken equipment to the repair shop where the equipment is sold.

The farm equipment mechanic sometimes works outdoors in bad weather. During an emergency his working hours can be long. Sometimes he works at night by floodlight to repair a machine in time for its job the next morning.

Who can be a farm mechanic? A person who

- likes to use tools and understands machinery
- knows how to use *welding, brazing,* and *blacksmith equipment*
- can use ordinary tools in new ways for special jobs
- is willing to work at any time of the day or night in all kinds of weather
- doesn't mind getting dirty.

V.I.P. WHO WORK WITH EARTH-MOVING MACHINES

Operating engineers run and repair the earth-moving machines used in the building or *construction industry.* Most of these machines are larger and more complicated than the equipment used on farms.

Operating engineers are V.I.P. who work on exciting jobs. All over the world, they clear land for houses for shopping centers, and freeways. They reshape the land to make dams, harbors, and airports.

Operating engineers usually specialize. They may become

UNIVERSAL EQUIPMENT OPERATORS,
GRADING AND PAVING OPERATORS,
PLANT EQUIPMENT OPERATORS, or
HEAVY DUTY REPAIRMEN.

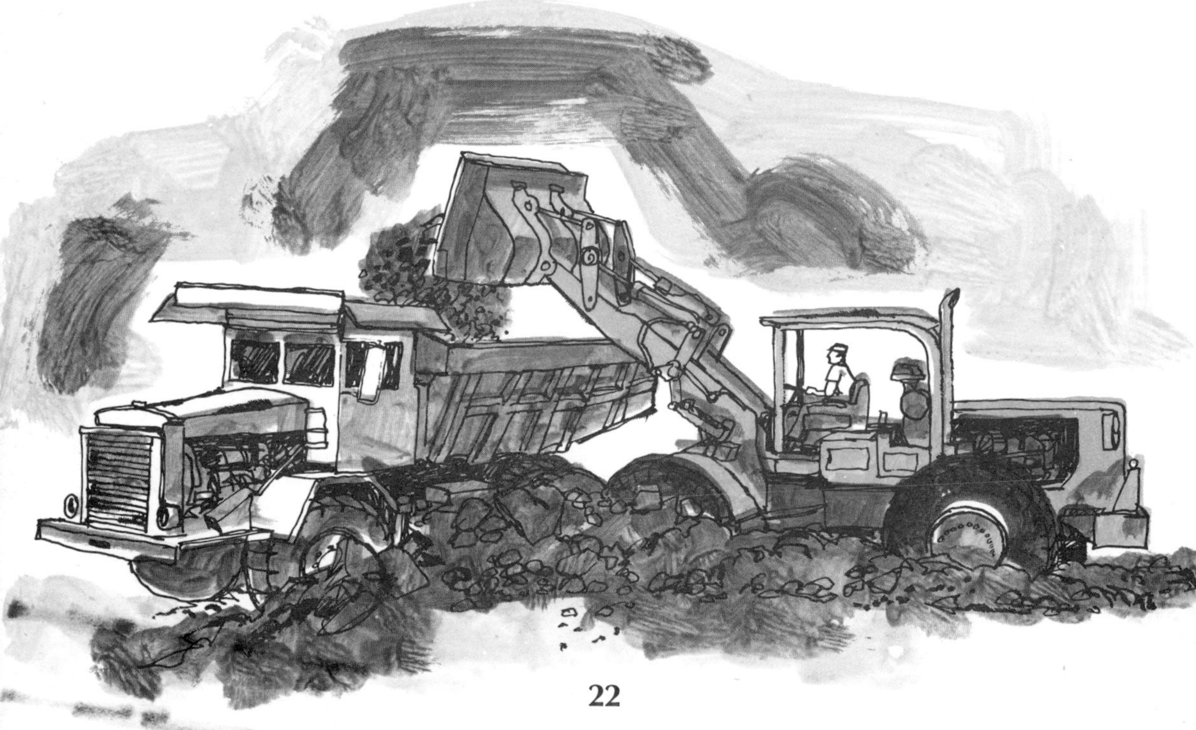

UNIVERSAL EQUIPMENT OPERATORS

Universal equipment operators work with large machines that dig, lift, haul, pull, push and hammer.

A few of these *vehicles* and machines are cranes, *loader-backhoes, power shovels, derricks,* and *pile drivers.*

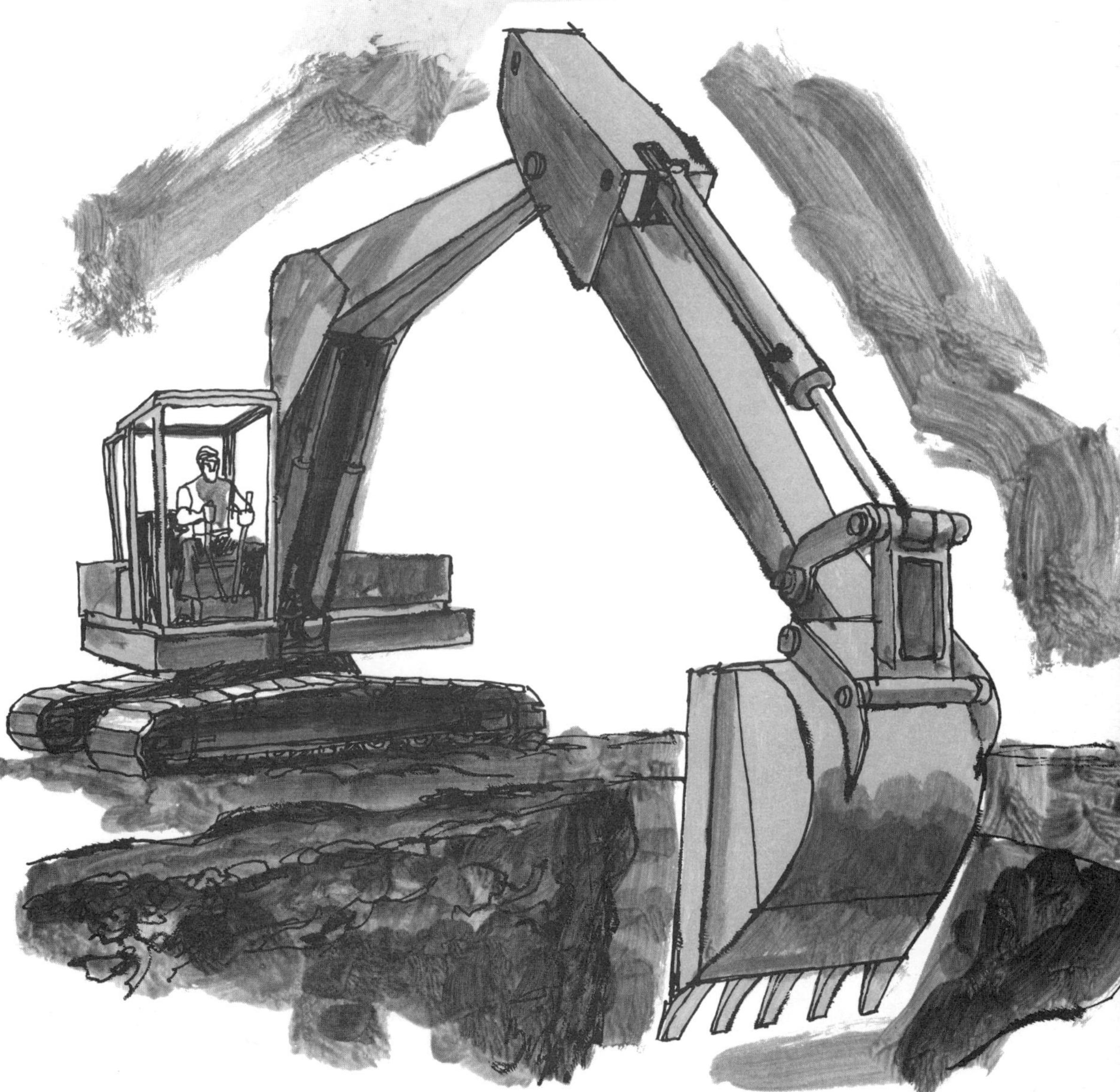

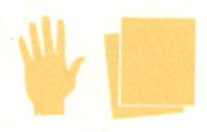

CRANE OPERATORS (cranemen) have some of the most difficult jobs.

Because a crane has many attachments, a craneman can do many kinds of work.

Buckets and clamps are used at the end of the *loadline* for lifting or digging.

A *platform attachment* can lift a load of lumber or even a group of men.

One crane operator can lift a concrete wall into place.

With a *wrecking ball* on the loadline of his crane, he can knock down an old building. Workers call the wrecking ball a *"skull cracker."*

As he works with his crane, the craneman moves pedals with his feet and levers with his hands. He rotates the crane on its *chassis* and raises and lowers the *crane boom*. The levers control the length of the loadline and move the attachments at the end of the loadline.

The craneman knows how much weight his machine can lift. He has to figure out the weight of the material he plans to lift. And he must remember many facts and figures when he operates his machines.

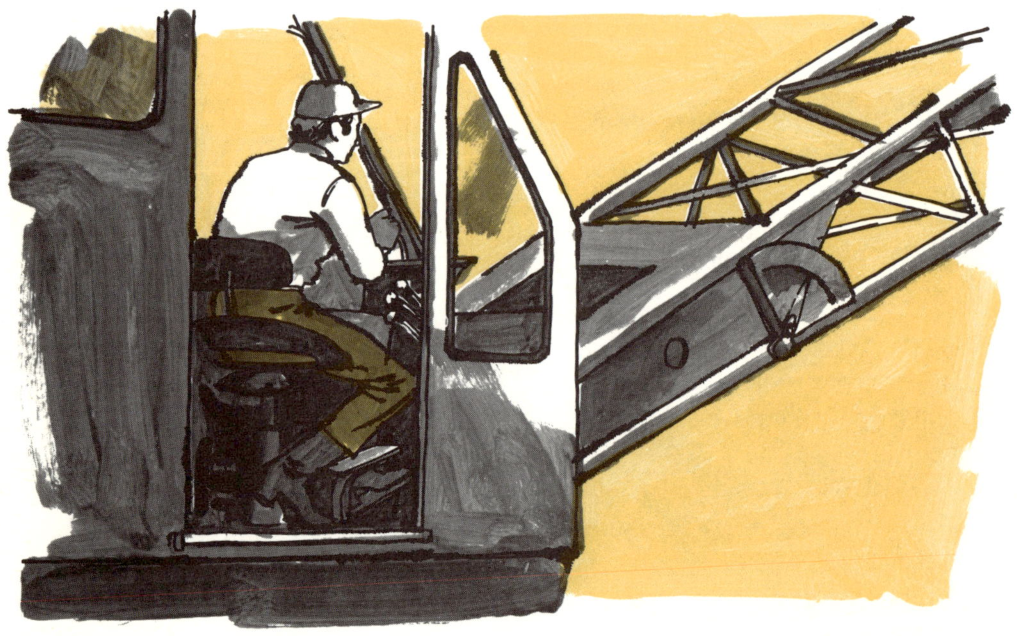

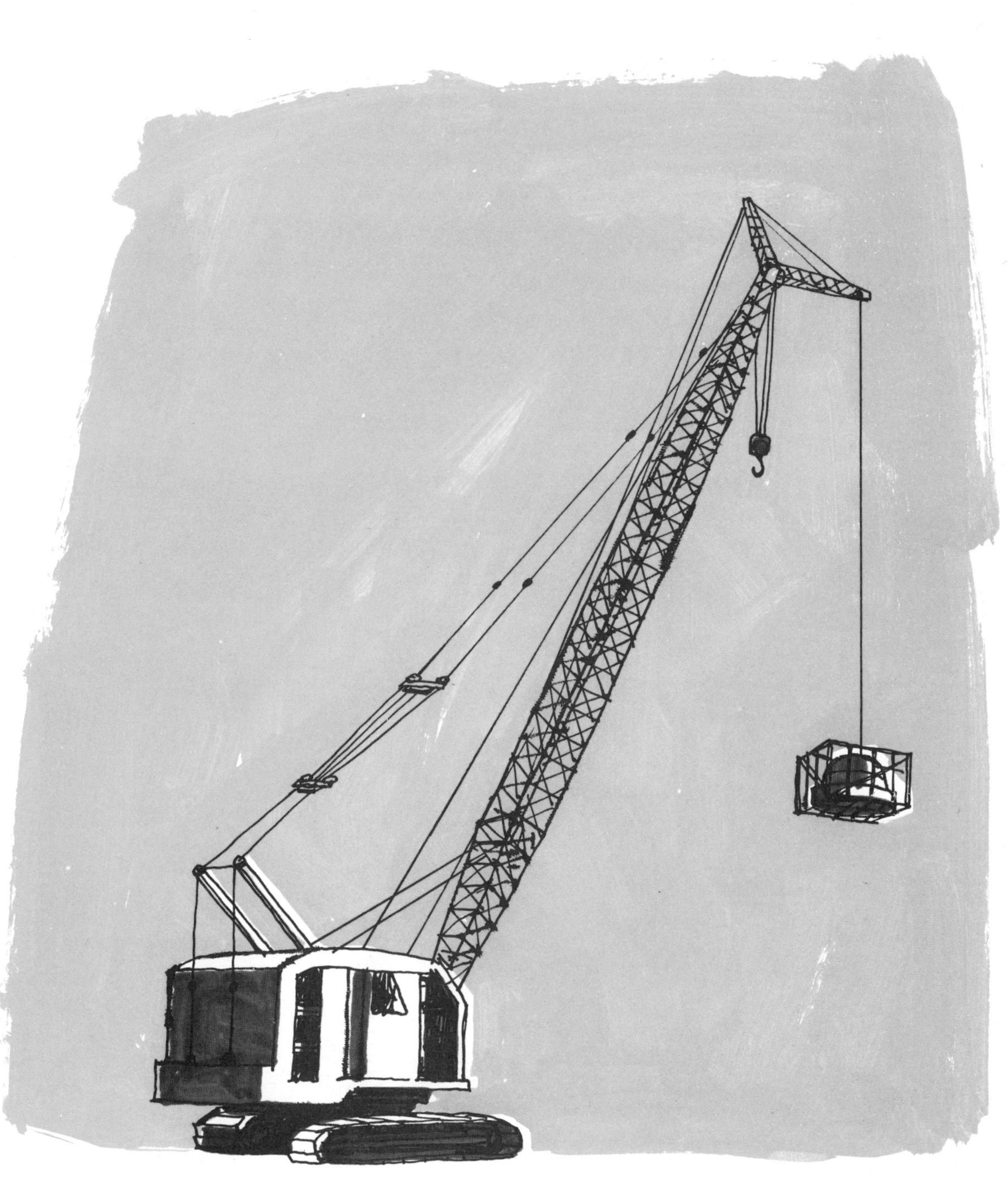

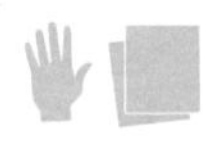

The **LOADER-BACKHOE OPERATOR** drives one of the most useful earth-moving machines.

The man and the machine can do many kinds of jobs. For instance, the backhoe can reach under the pavement of a road to *excavate* a tunnel. It can dig a square hole a yard wide and 20 feet deep.

The machine has a bucket-like scoop at the end of a sturdy mechanical arm. This is built onto the front or back of a tractor.

When the operator wants the bucket to dig, he turns a dial to "dig" and pulls a lever. The bucket moves forward, scoops backwards, and fills itself. He turns the dial again, and the bucket twists sideways to empty itself.

If the operator dials "trench" and pulls another lever, the bucket digs a trench and levels the bottom of it.

The loader-backhoe operator has to be careful not to cut through any gas or water pipes, or electrical *cables*. He must be able to read charts that show where pipes and cables are buried.

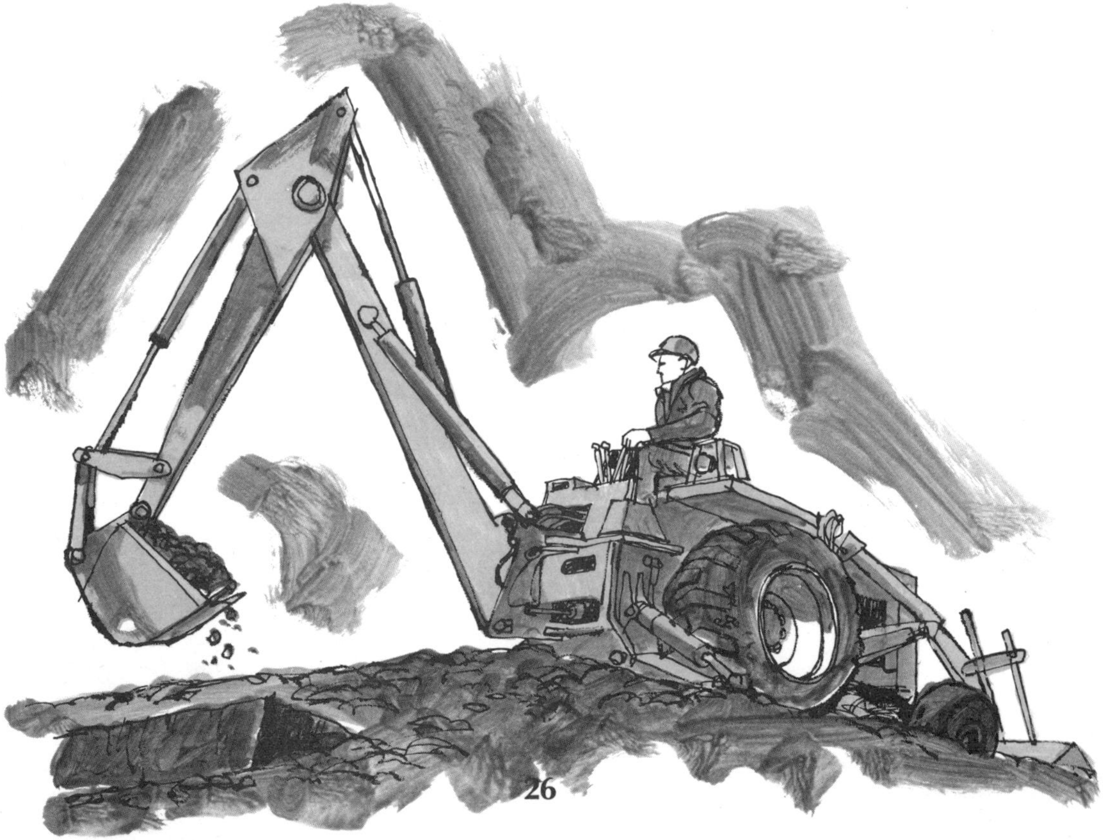

A power shovel is a large scoop at the end of a long, mechanical arm. The **POWER SHOVEL OPERATOR** in the cab can move the scoop up and down and from side to side. He can scoop earth from the hillside or reach down to dig in a deep hole.

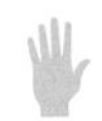

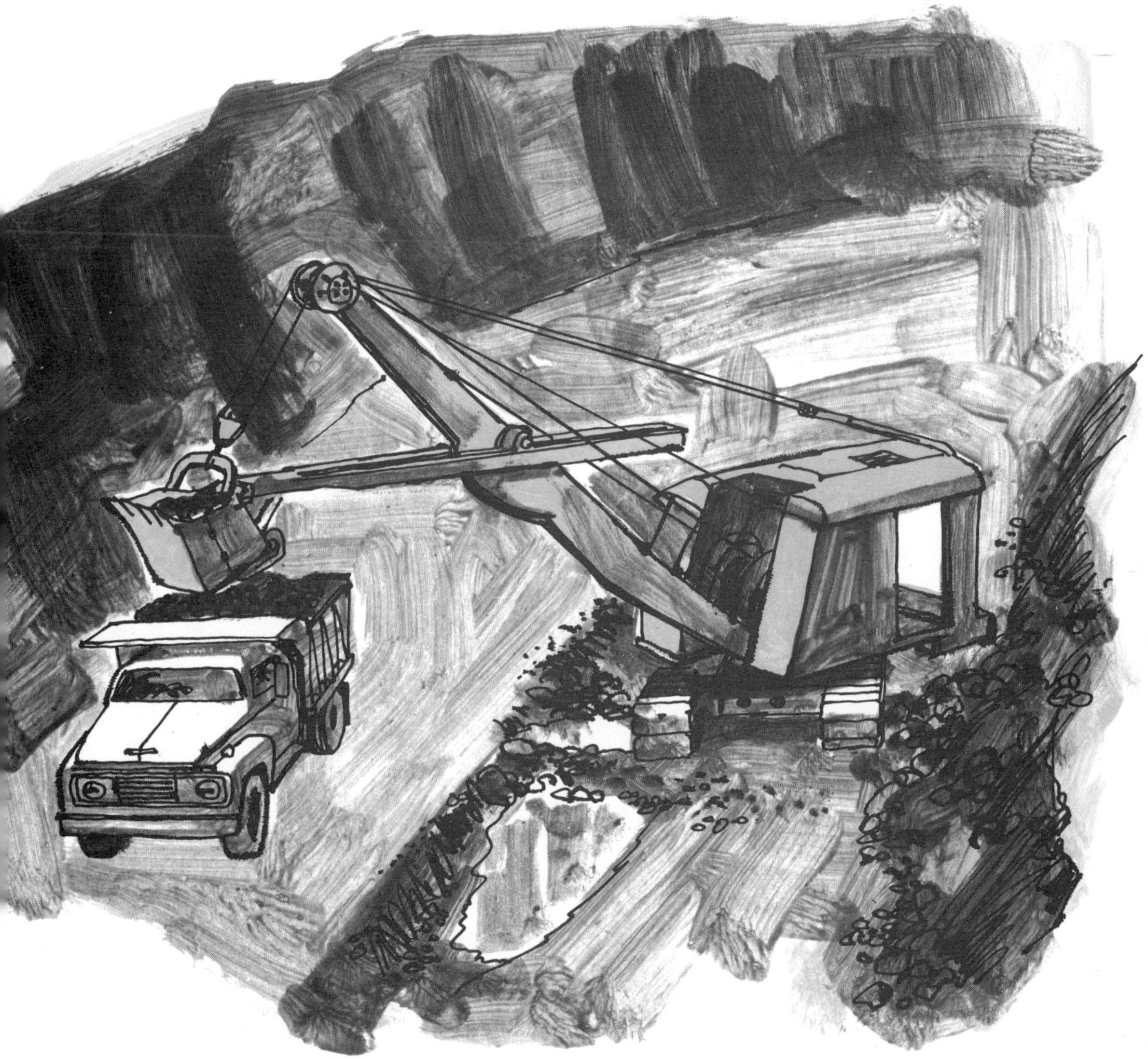

A derrick is a hoisting or lifting machine. It is like a crane, but a derrick usually stands in one place and moves heavy loads from side to side. The **DERRICK OPERATOR** lifts *cargo* from the deck of a ship and sets the load down on the dock.

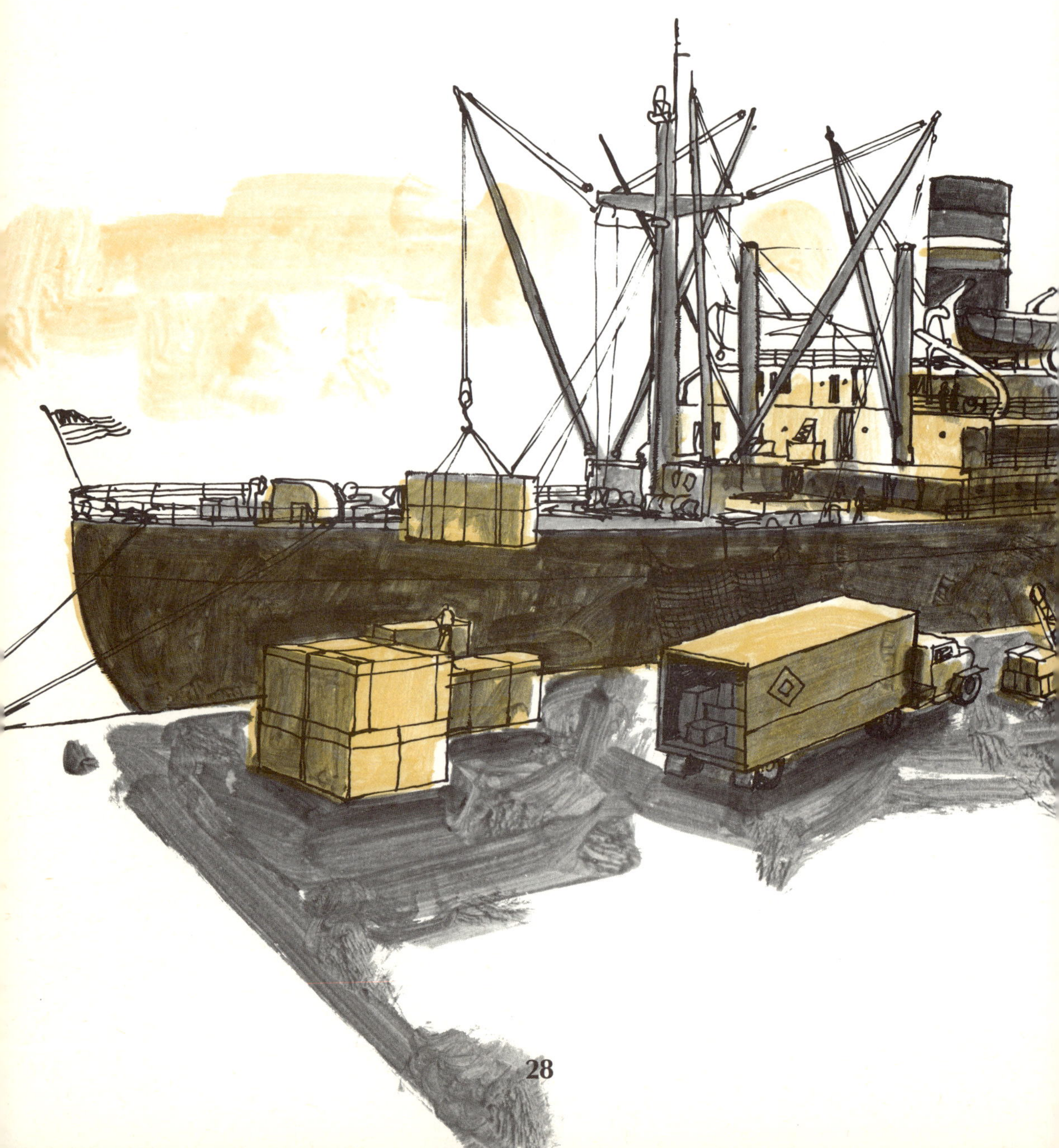

A pile driver is a very powerful and noisy machine. The **PILE DRIVER OPERATOR** uses it to drive wood or steel posts, called piles, into the earth. The piles help support buildings, bridges, or freeways.

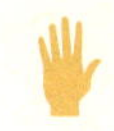

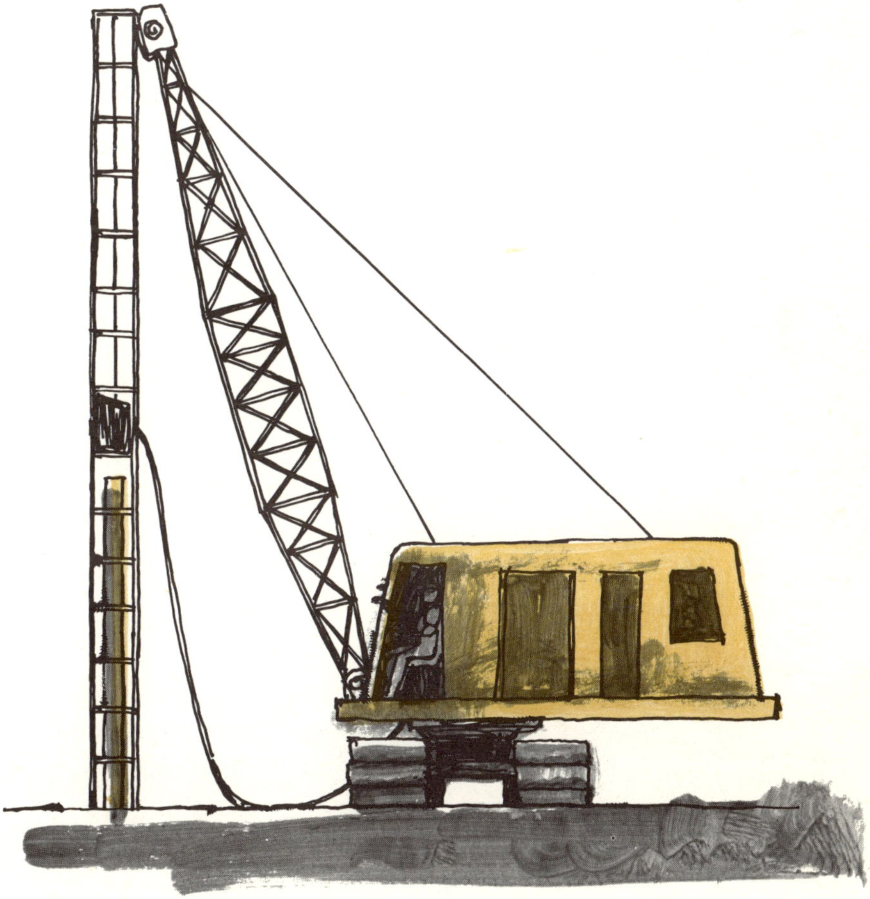

GRADING AND PAVING OPERATORS

Grading and paving operators work with machines that build roads and highways. Grading is preparing the land for a road. Paving is covering the surface of the road with *concrete* or *asphalt*.

These V.I.P. can operate many different machines. Some of the machines and vehicles are *bulldozers, rollers, earth borers, scrapers, graders,* and *concrete mixer-pavers.*

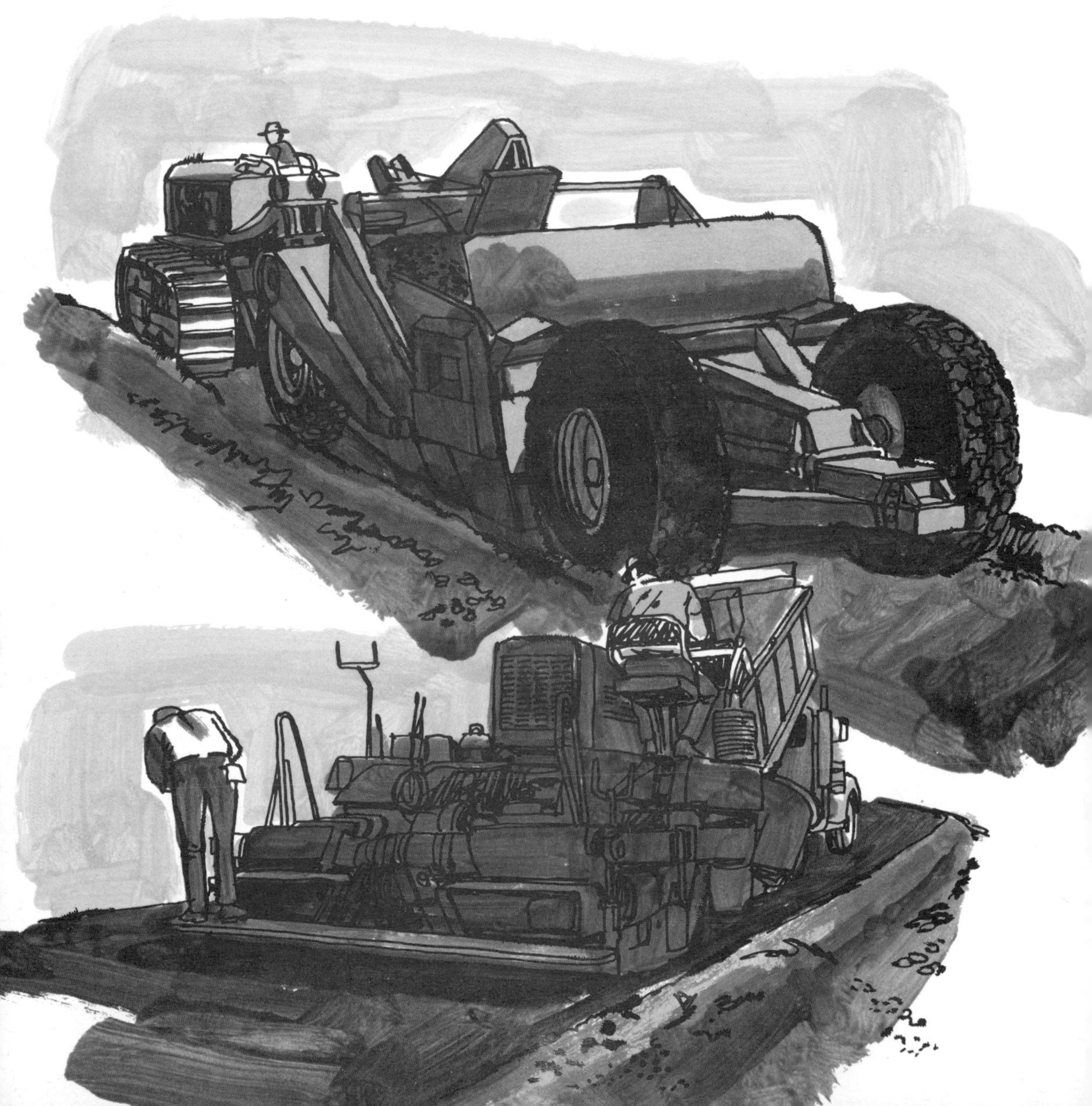

The **BULLDOZER OPERATOR** drives a tractor fitted with 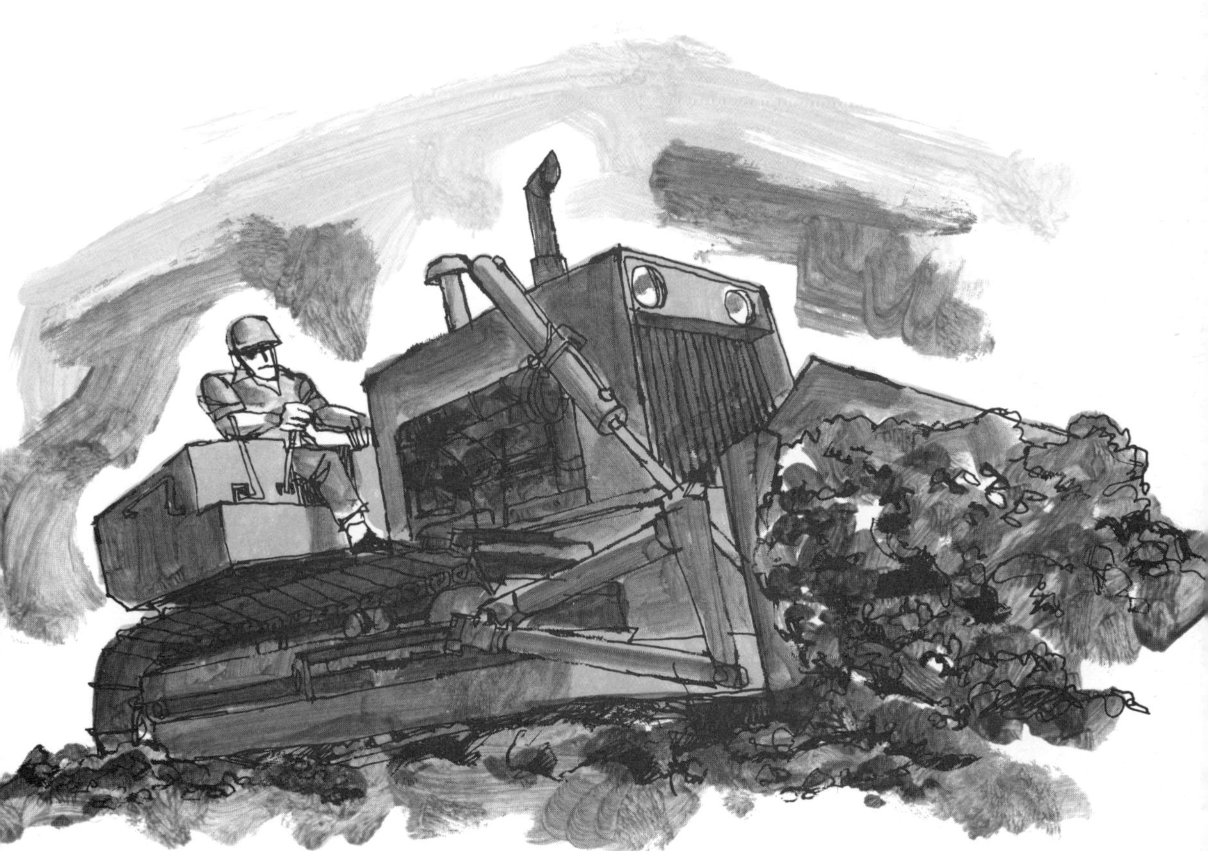 a large blade. He first pushes the earth into piles, then spreads the piles evenly.

For his own safety and that of his equipment, the operator must know exactly what his dozer can do.

He must know how steep a *grade,* or hill, his machine can go up or down safely. If he uses the wrong speed and gear, he can ruin the equipment, and he could seriously injure himself or others.

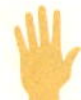

ROLLER OPERATORS drive *flat wheel rollers* and *sheep's foot rollers* which pack down the earth to make it more solid. The operator knows how water drains onto or away from the earth he is flattening. He uses this knowledge to help him make a roadbed that will not wash away or sink after heavy rains.

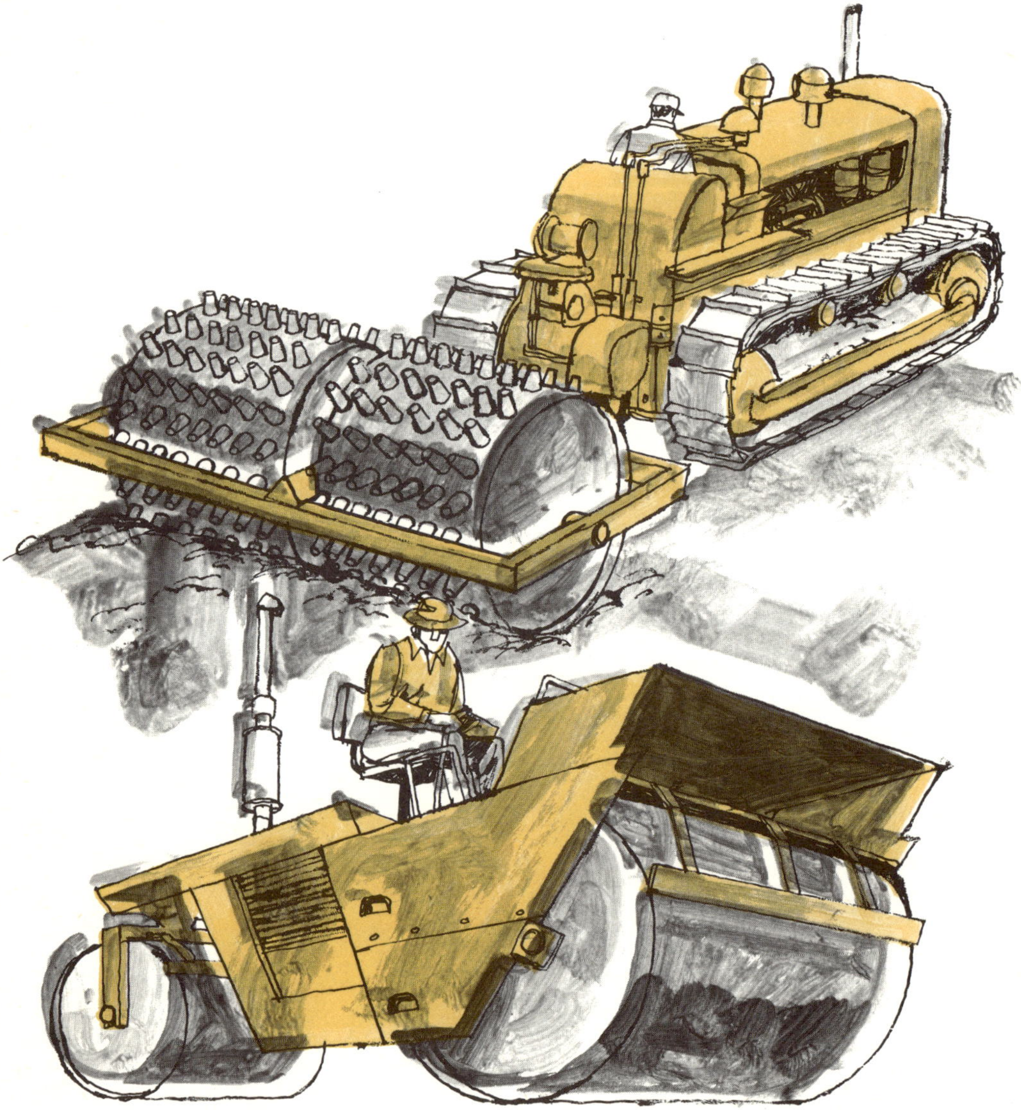

The **EARTH BORER OPERATOR** drills holes for posts and other supports.

He selects the right *auger* (drill) for the size hole he needs. He sets the auger in the *spindle* which holds the auger as it turns.

This job is fairly easy to learn. The operator starts the machine and stops it when a gage tells him the hole is the right depth.

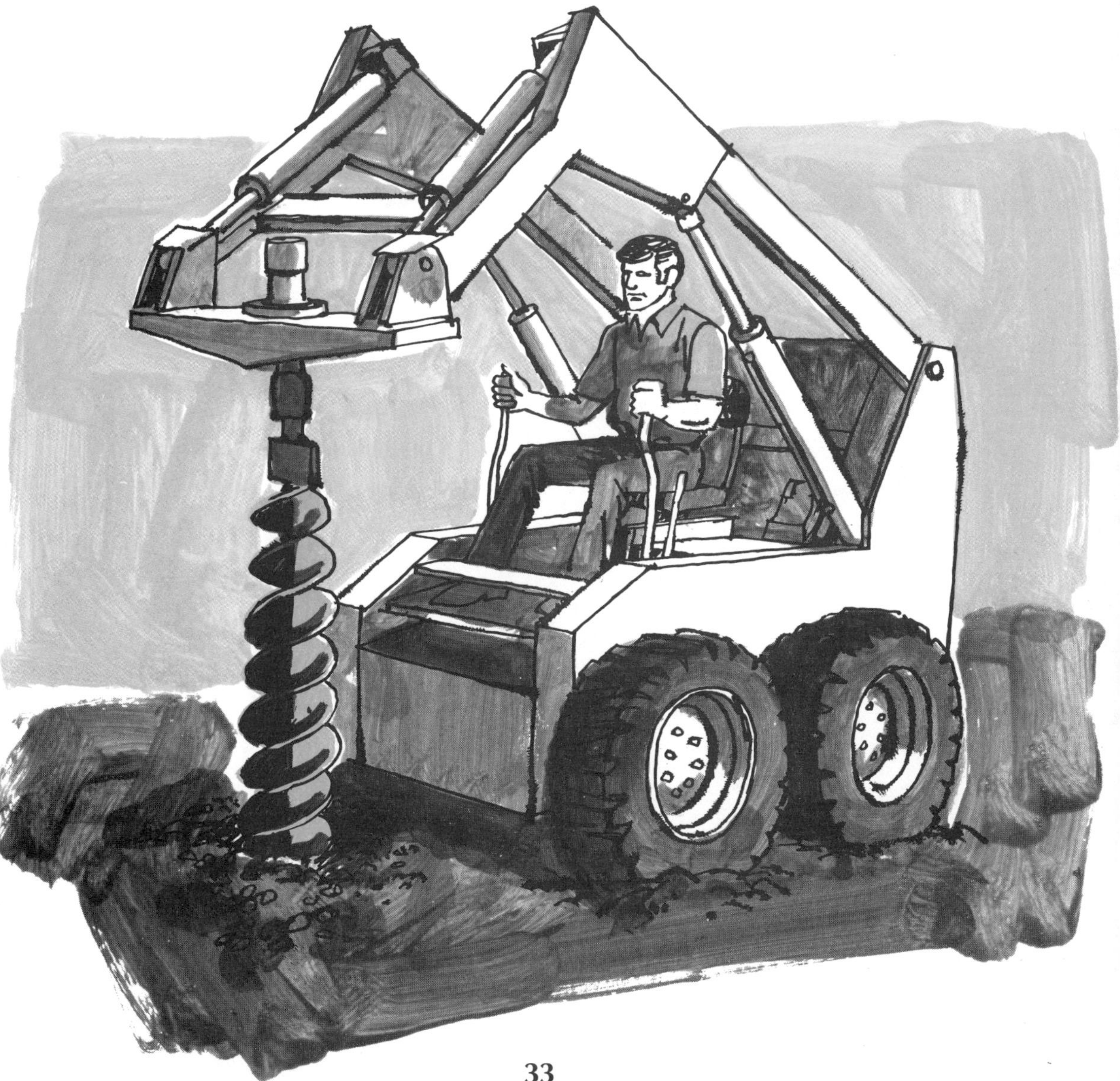

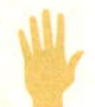

The **SCRAPER OPERATOR** pulls the scraper with a powerful tractor. The scraper is a wide scoop. It quickly picks up tons of earth from the surface to clear a space for a road or an airport.

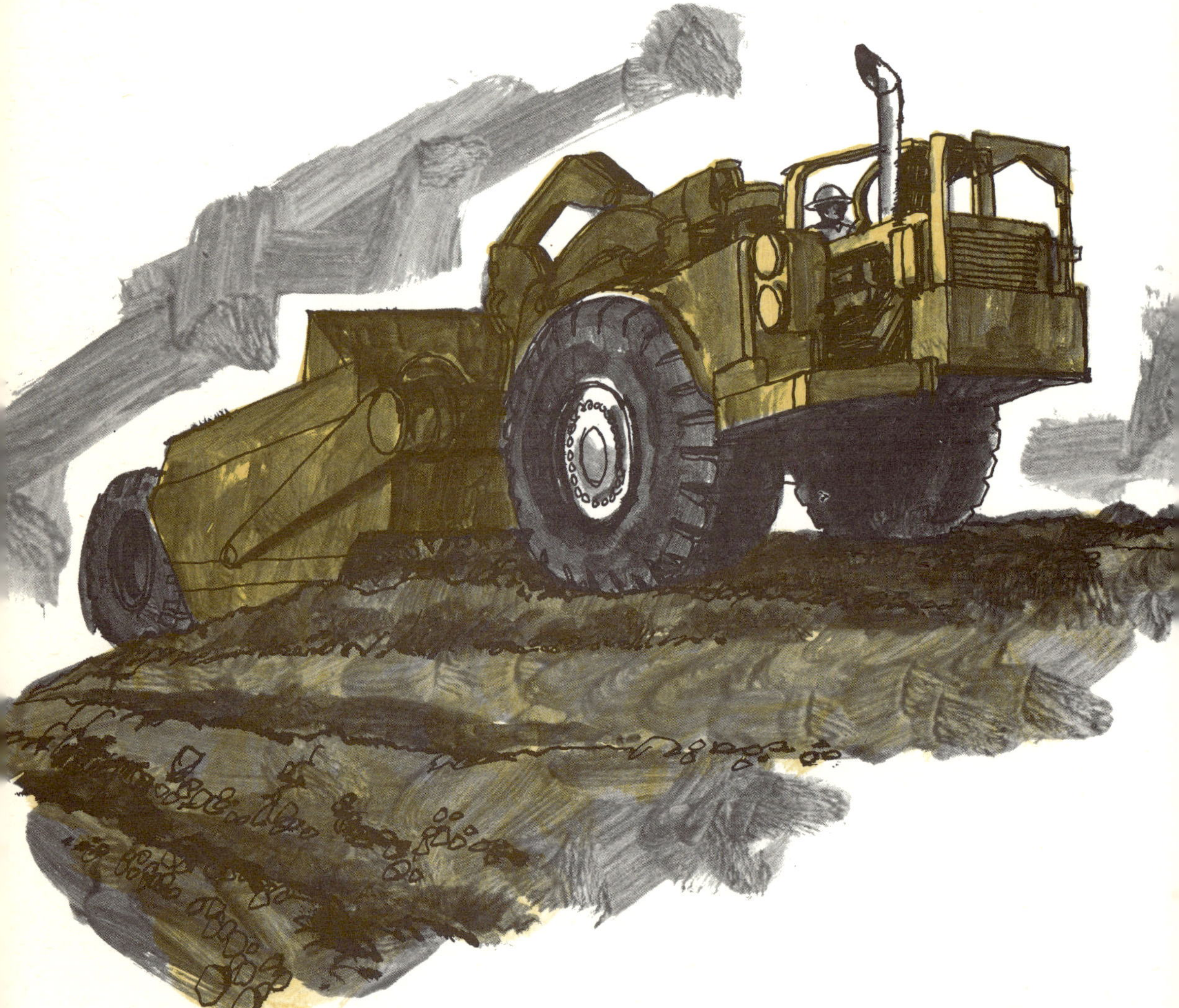

A grader is a tractor with a large blade fastened crosswise between the front and back wheels. With his control levers, the **GRADER OPERATOR** in the cab can lift the blade and turn or slant it to smooth and level the earth.

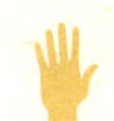

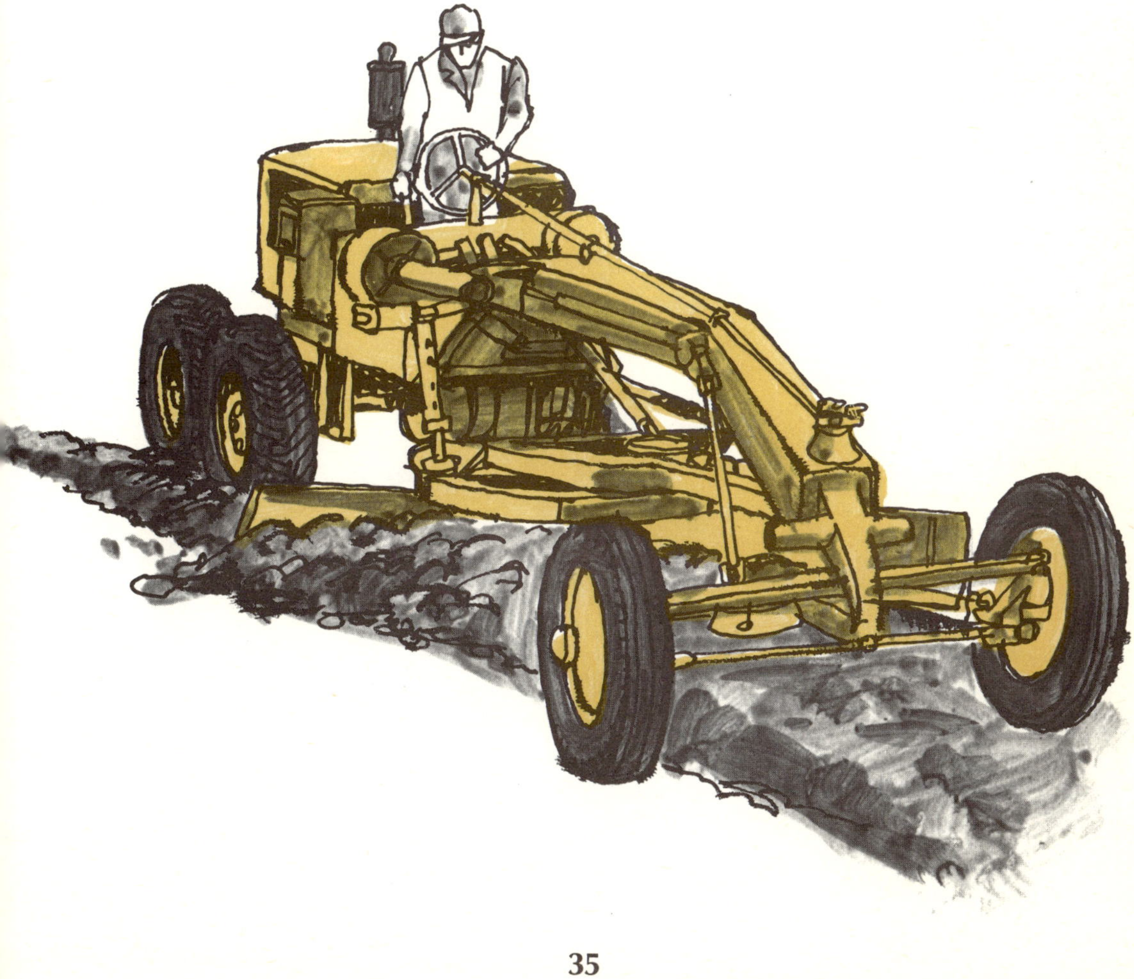

Several V.I.P. working together operate the concrete mixer-pavers. The machines follow each other in a line.

The first operator mixes and pours the concrete from his machine onto the roadbed.

The next operator's machine is a *spreader* which levels the concrete.

Another operator smooths the wet concrete with his machine.

Workers with hand tools roughen the surface of the concrete so that it will not be slippery for cars that travel on the finished highway.

The last workers at the end of the concrete-paver line cover the surface with large sheets of paper so the concrete will not dry too quickly.

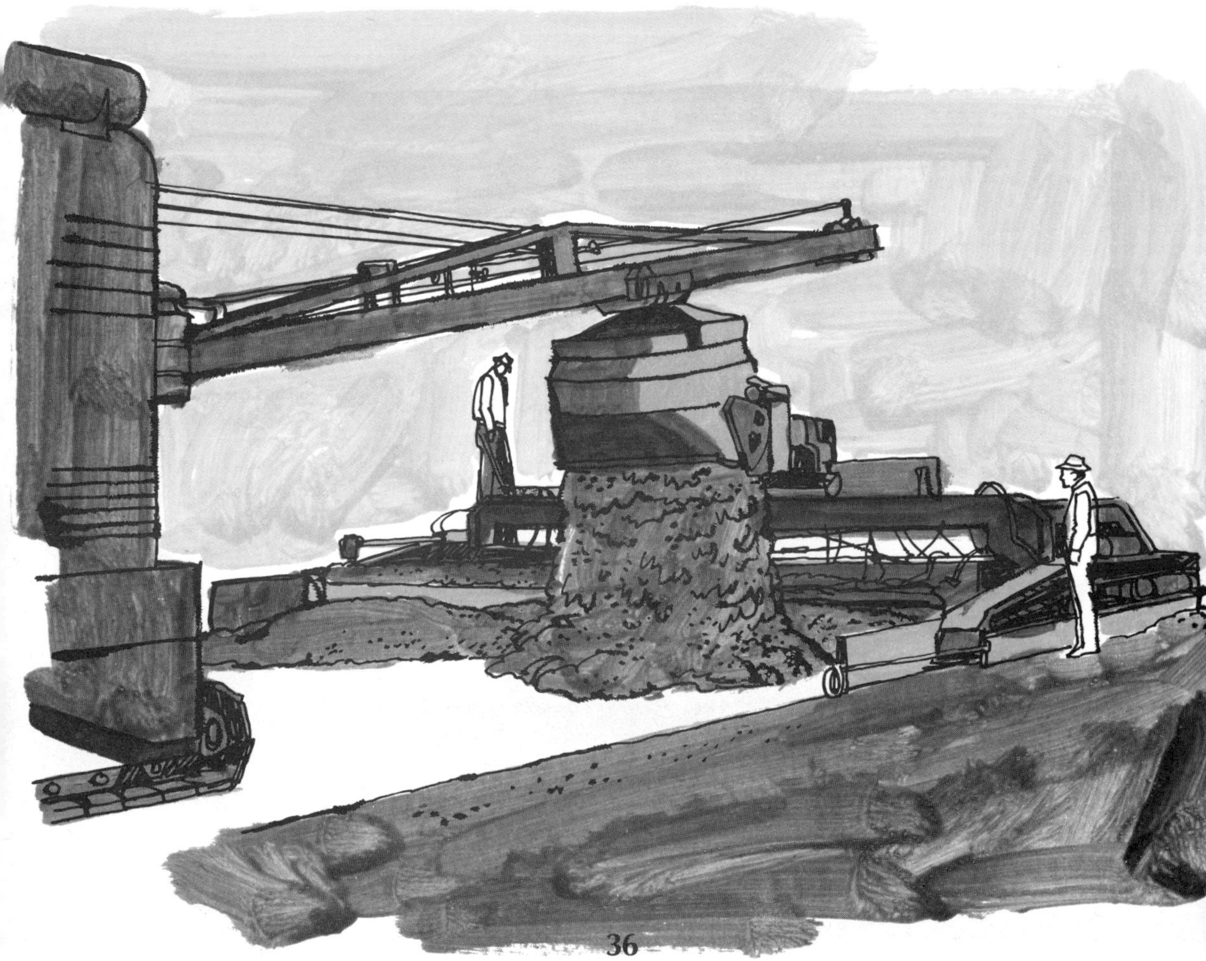

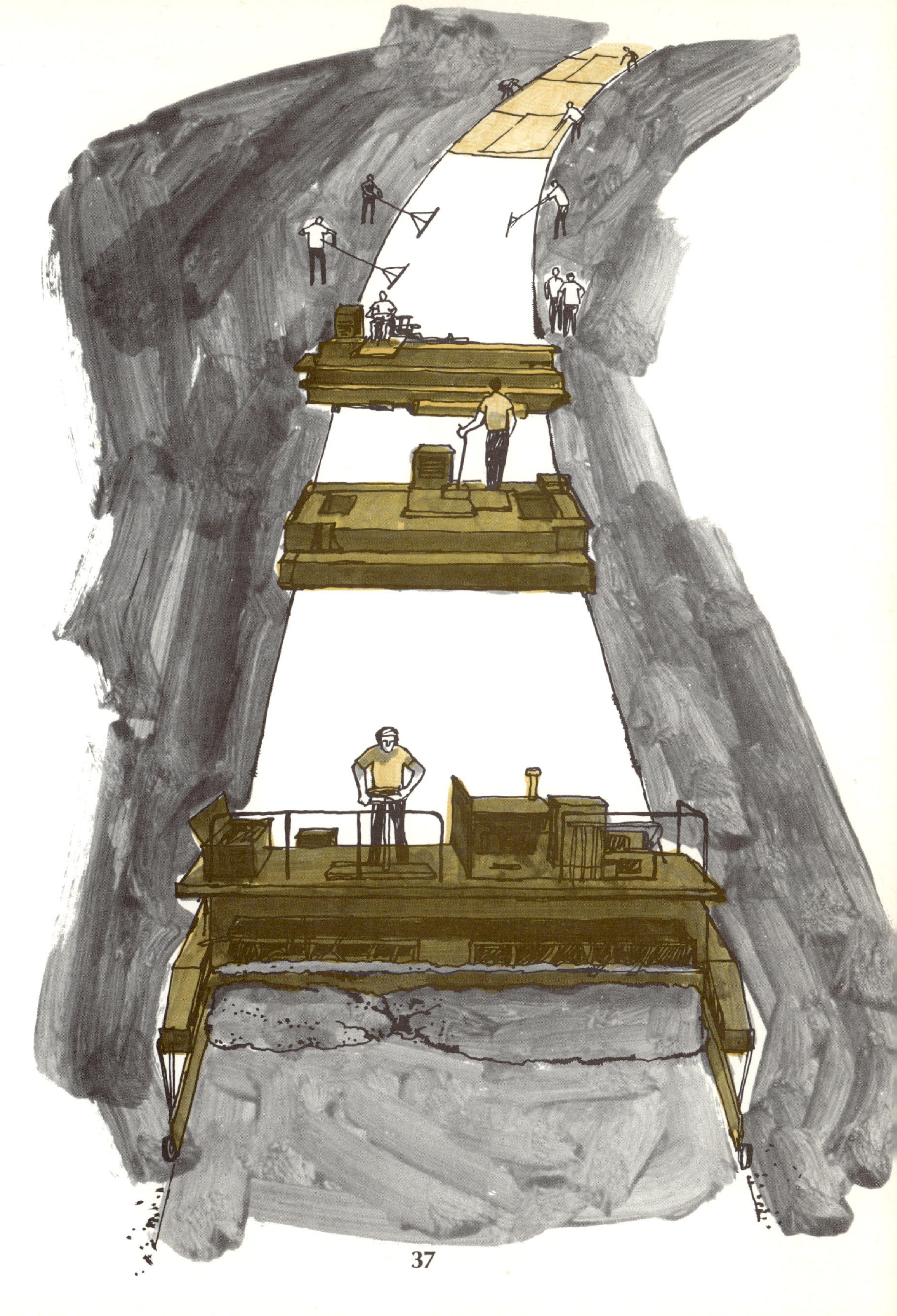

PLANT EQUIPMENT OPERATORS

Asphalt is one of many different materials used to build roads and highways. Sand, gravel, stone, and cement are some of the others.

When you see a big truck on its way to a road *construction site,* you may wonder where the driver got his load.

All road-building materials come from a place where workers operate machines that sort, crush, mix, heat, and load these materials.

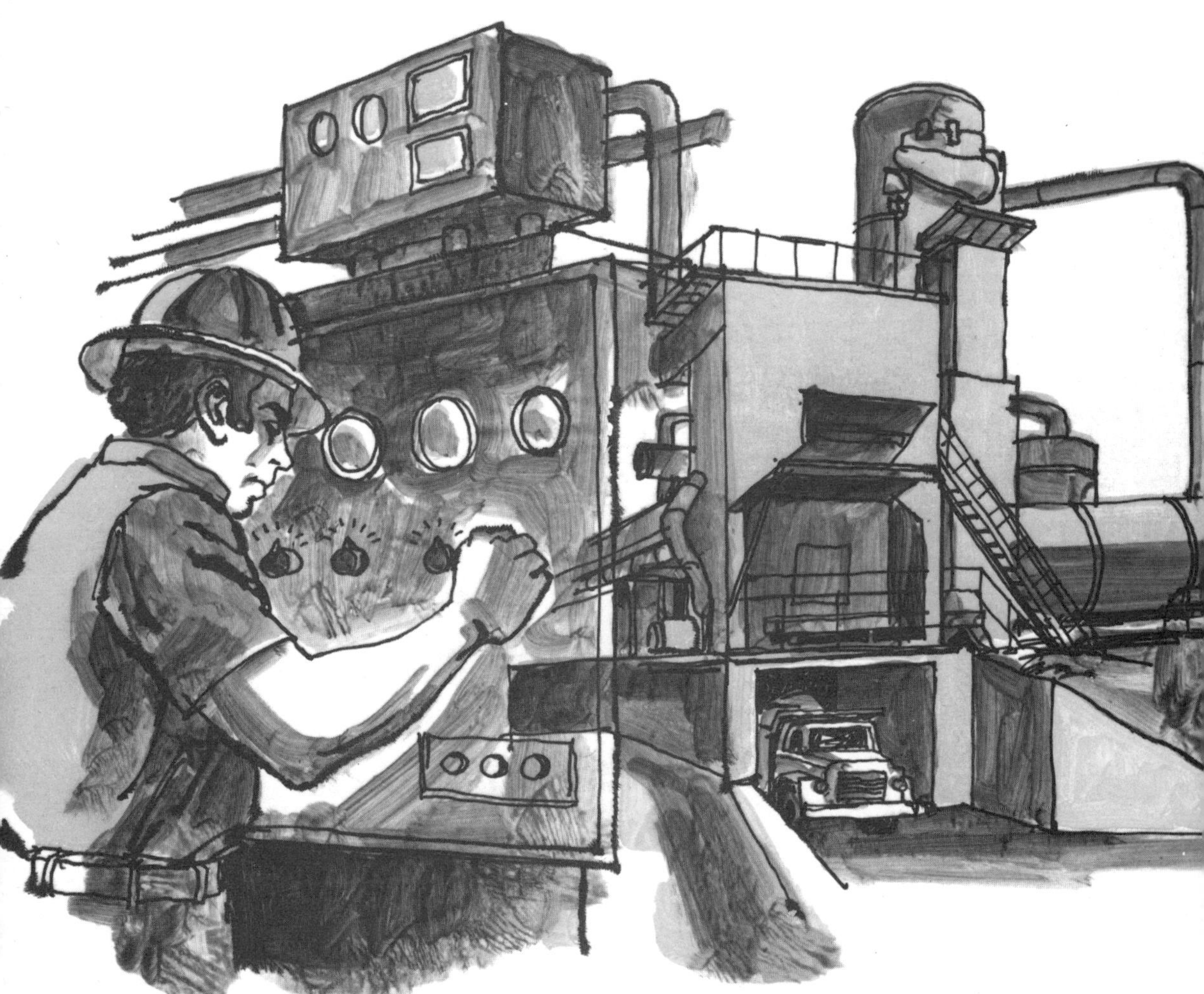

The V.I.P. who operate the machines at this place, called a *batch plant,* are **PLANT EQUIPMENT OPERATORS.** They keep fires at the right temperature to heat and dry materials. They weigh sand and gravel, then mix them with cement in a *mixing drum* to make concrete.

The plant equipment worker uses many machines.

One machine is a conveyor belt that carries rock materials to a *crusher.*

Another is a motor-powered *shaker screen* that sorts rocks by letting pieces of a certain size pass through the screen.

The plant equipment operator loads trucks that carry paving materials.

Sometimes he loads sand from an overhead storage bin called a *hopper.*

When materials, such as crushed rock, are in piles on the ground, he scoops them up with a *front-end loader* attached to a tractor.

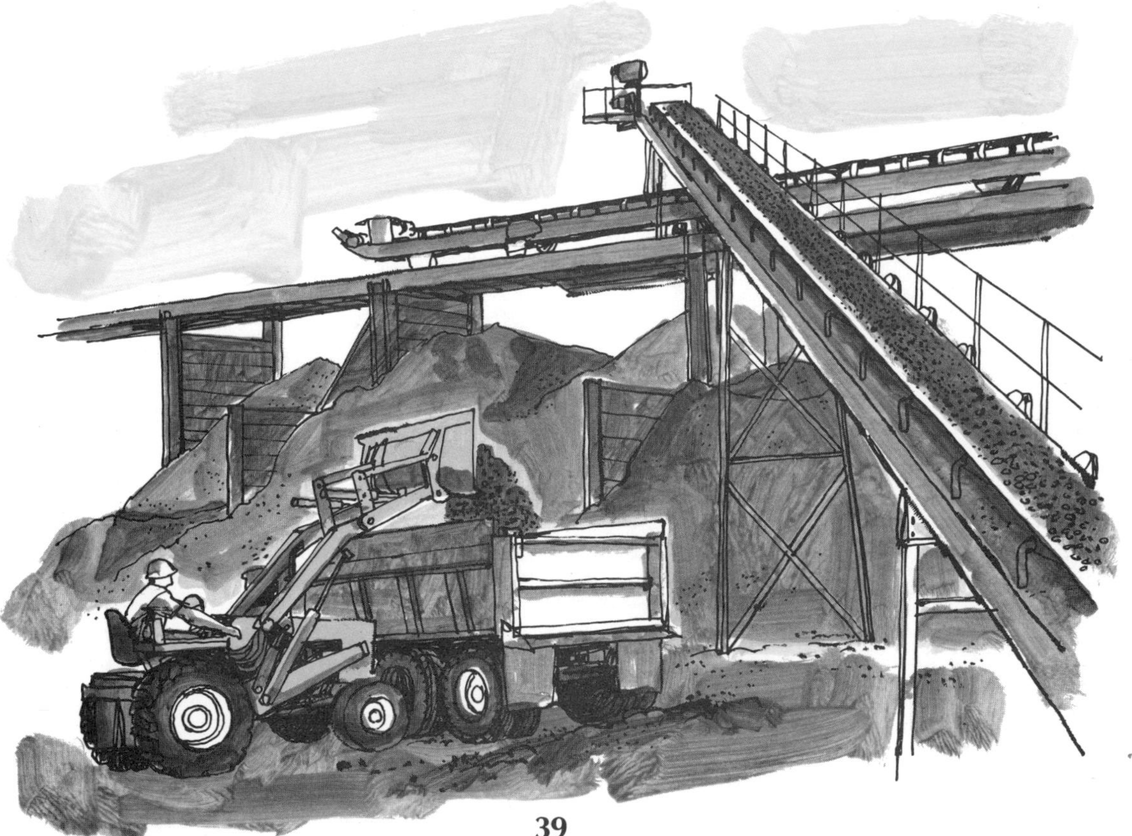

HEAVY DUTY REPAIRMEN

Repairmen with special training do the big repair jobs on earth-moving vehicles and attachments.

These men are **HEAVY DUTY REPAIRMEN**. They specialize in this work from the beginning of their training. They learn how to operate the construction equipment and how to repair it.

Whenever possible, these mechanics make repairs at the construction site.

If a machine can't be repaired at the construction site, it is taken to an equipment dealer's garage. Mechanics at this garage specialize in repairing the brand of equipment the dealer sells.

These mechanics have a good place to work. The garage has a high ceiling so large vehicles can be brought indoors. They use *hoists* and *overhead pulleys* to lift huge parts.

It takes many years of training and experience to be a skilled heavy duty repairman. These V.I.P. keep the machinery working so operators can finish their jobs on time.

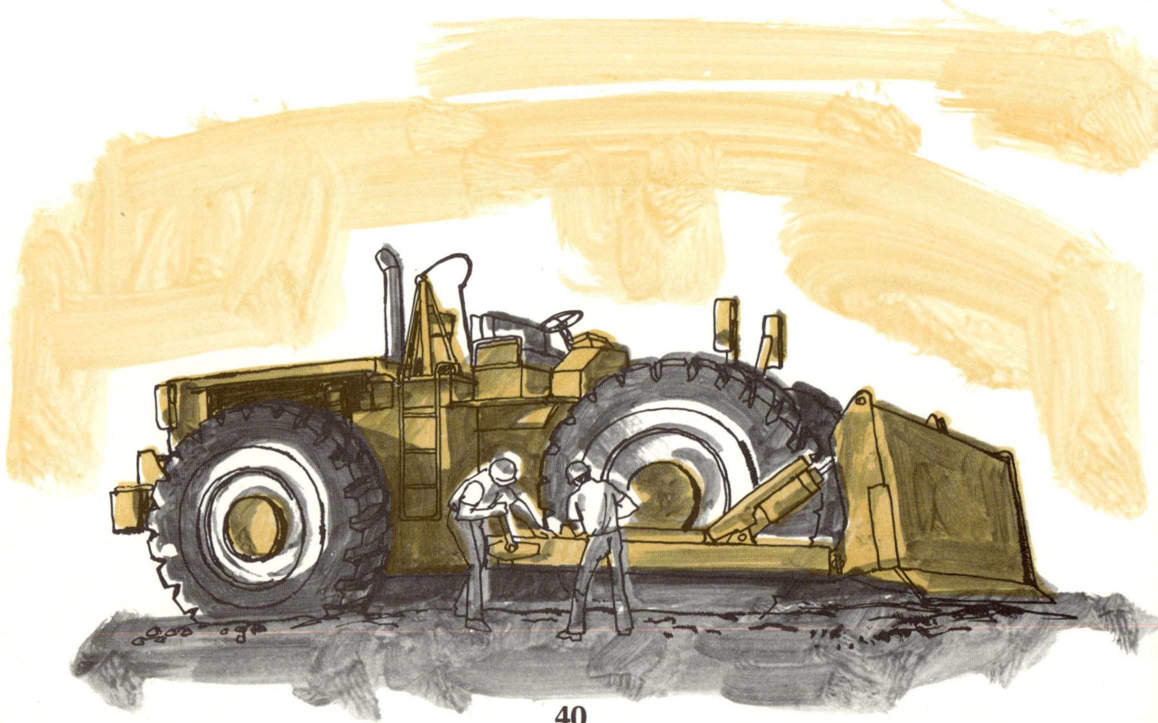

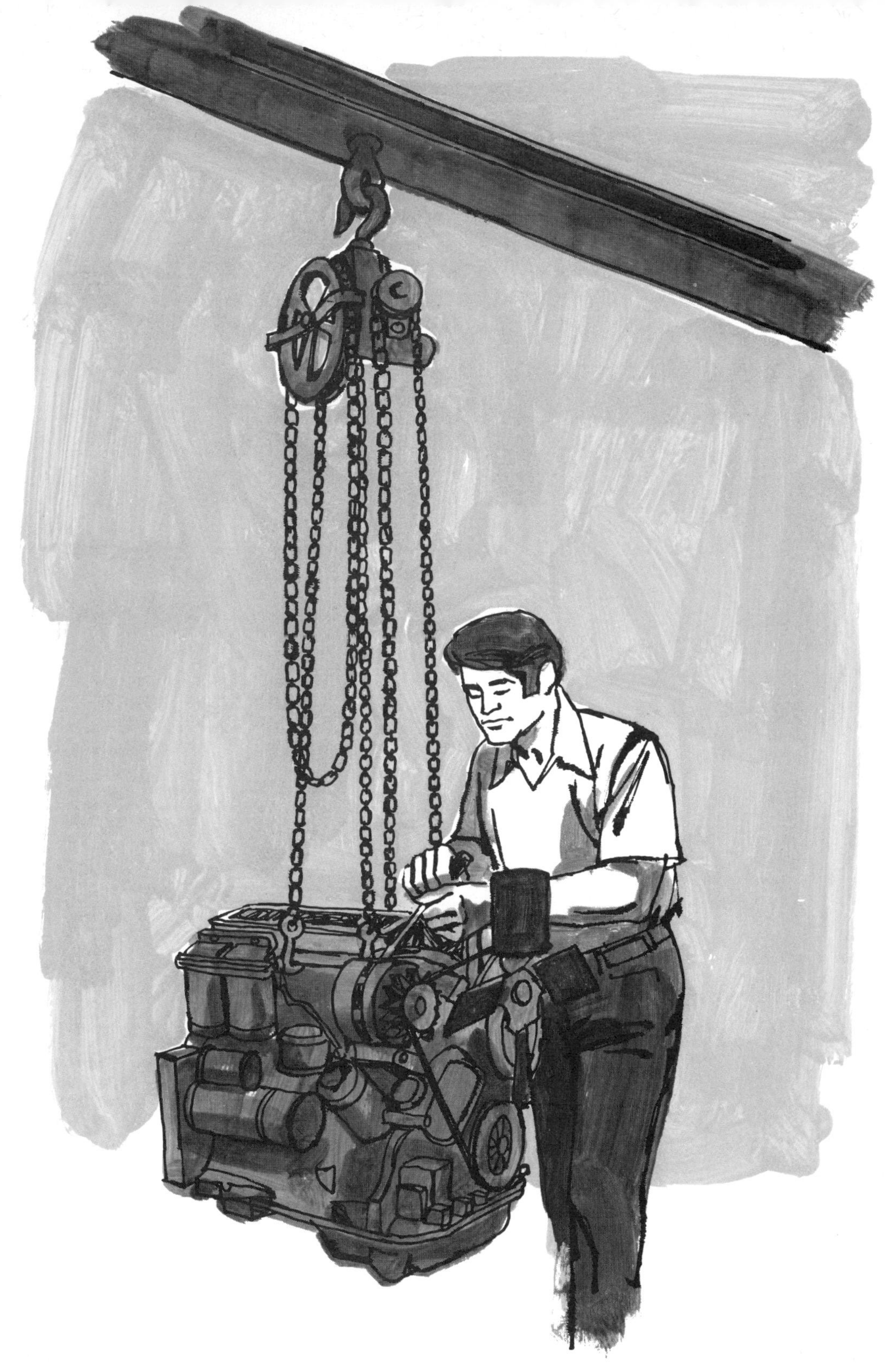

TRAINING FOR OPERATORS OF EARTH-MOVING MACHINES

There is a *union* for operating engineers who run the earth-moving machines. It's called the *International Union of Operating Engineers*.

A union is an organized group of workers. Members form committees to plan training, wages, and work hours for all union members.

UNION APPRENTICESHIP PROGRAM

The union offers training which young people can begin when they are teenagers. They start out as *trainees,* called *apprentices.*

Each apprentice signs an agreement to train for at least three years. He receives some pay during this time.

An apprentice learns on the job. He has a supervisor who checks his progress and helps him learn the job.

The apprentice gets about 6,000 hours of training. He spends 430 of these hours in classes given in high schools or *vocational schools.*

During his on-the-job training, the apprentice may work for several different contractors. This way, he gets experience operating machines on many different kinds of jobs.

Sometimes apprentices, who have had some experience with farm or construction machines in the armed services, get credit for their experience. They may not have to spend as many hours in the apprentice training program.

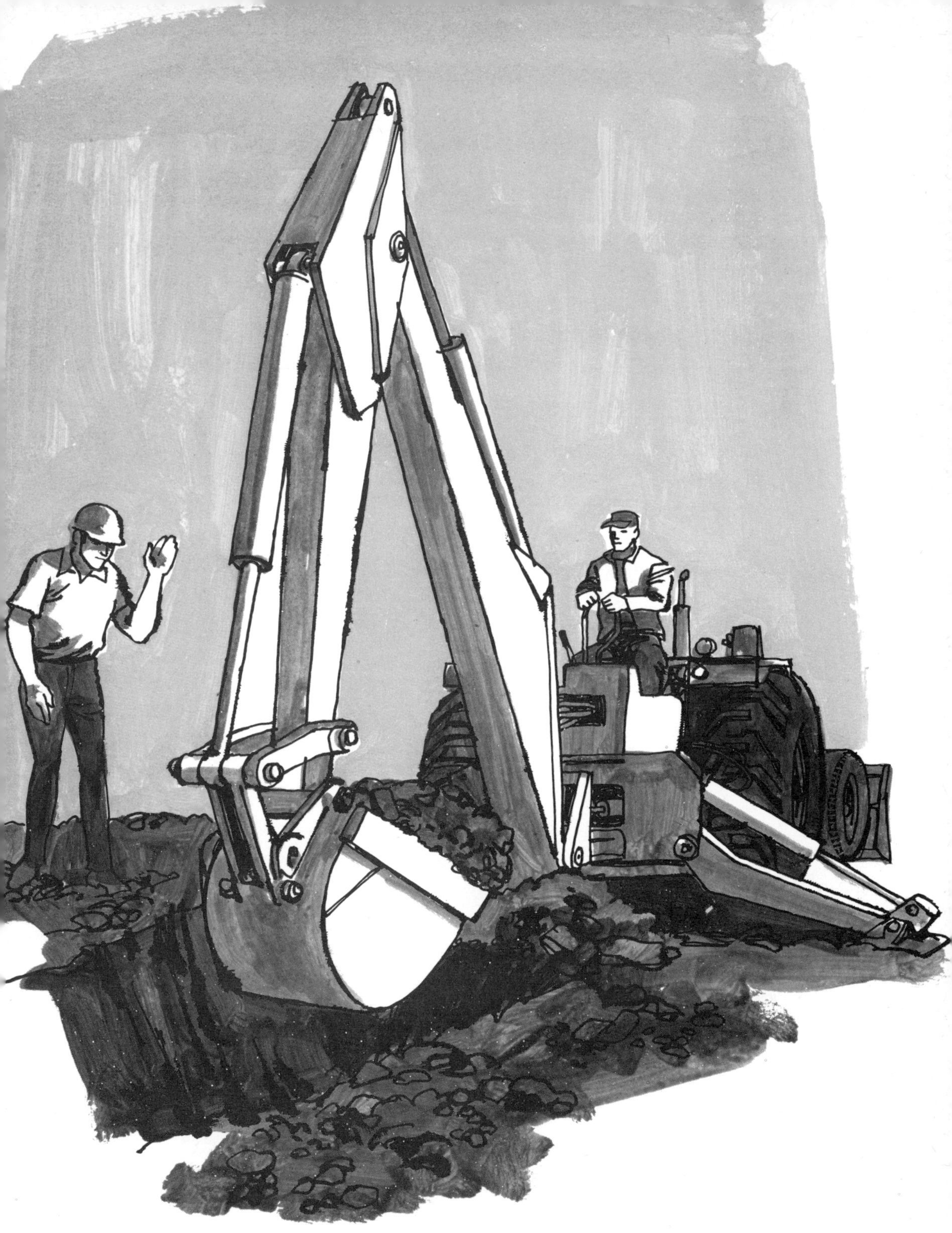

PRE-APPRENTICESHIP TRAINING

The operating engineers' union also has *pre-apprenticeship training* programs. These courses, which are held on the job, last from six to twelve weeks.

To train as bulldozer operators, the students practice moving dirt around with a bulldozer. They pile it high and cut through the piles to test their skill in handling the big machines.

In New York, a group of pre-apprentices did actual construction work using several machines. They built a lake, a ski slope, and a toboggan run for the Boy Scouts.

Pre-apprentices may earn a small amount of money while they are learning. If they work away from home, they usually get their meals and a place to sleep.

After their pre-apprenticeship training, the students who show they are capable and dependable are selected to join an apprentice program.

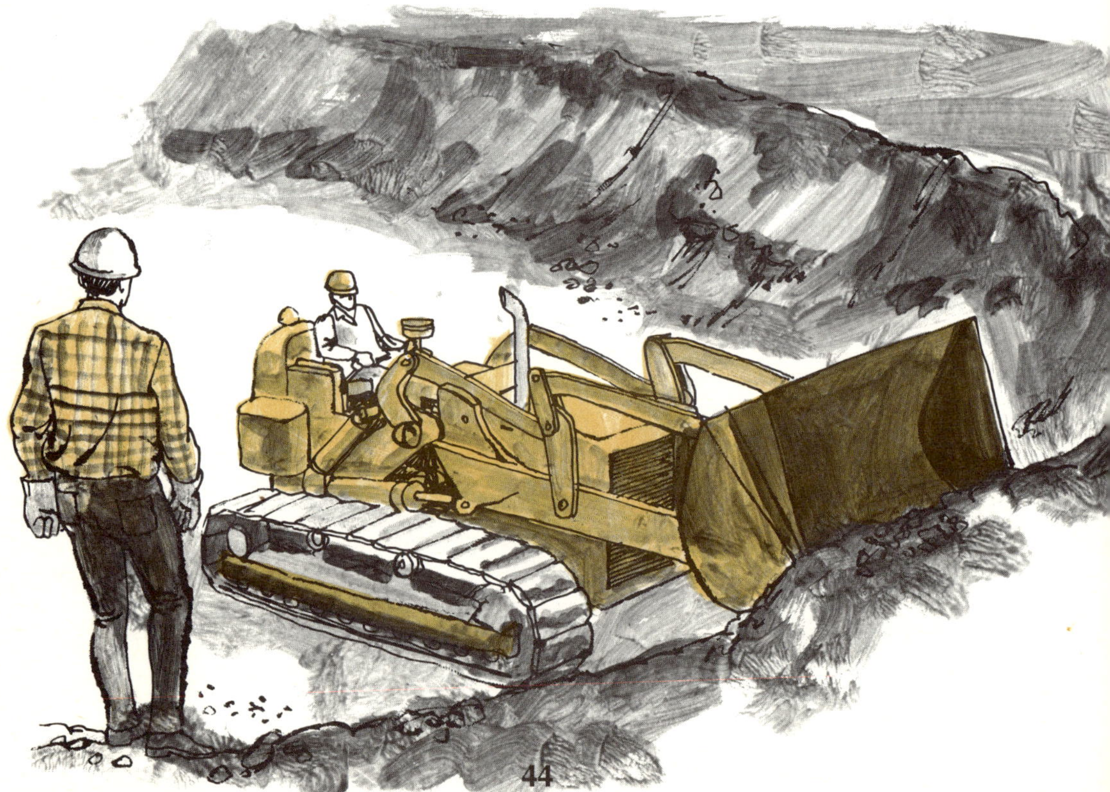

OTHER WAYS TO LEARN TO OPERATE
EARTH-MOVING MACHINES

You may wonder if it's possible to become an operating engineer without going through the union's apprenticeship program.

Yes, it is. If you get a chance to start out as an **OILER,** the person who oils the machinery, you could work your way up to operator or mechanic, but it takes a lot longer. 

Another way to get training is to go to a private school that specializes in teaching how to operate and repair heavy duty equipment. People who attend these schools pay for their training.

Who can be an operator of earth-moving equipment? A person who

- is physically strong and in good health
- can hear and see well
- can judge distances
- has good eye-hand-leg coordination
- has no allergies to dust or wild plants
- is not afraid of high places
- can act quickly with good judgment in unexpected situations
- understands and obeys safety rules
- has good mechanical skills
- is able to give first aid
- has finished high school or has equivalent education
- has had training in operating the machines.

Who can be a union operating engineer? A person who

- has all the above qualifications and
- is an American citizen, or is becoming one
- is accepted by the International Union of Operating Engineers as an apprentice
- is between 18 and 25 years of age.

V.I.P. WHO WORK WITH SPECIAL EARTH-MOVING MACHINES

Some operators work with equipment that does special jobs.

Earth-moving machine operators can lay pipe or cable in the earth. A skilled operator, with his machine, digs a trench, lays cable in it and covers the trench, all in one operation.

These operators work directly for telephone, electric, gas, and cable TV companies, or for *contractors* hired by these companies.

In places where it is difficult for operating engineers to go with their machines, **HELICOPTER PILOTS** help. They can lift huge telephone poles and set them in place. They can carry heavy loads to high places.

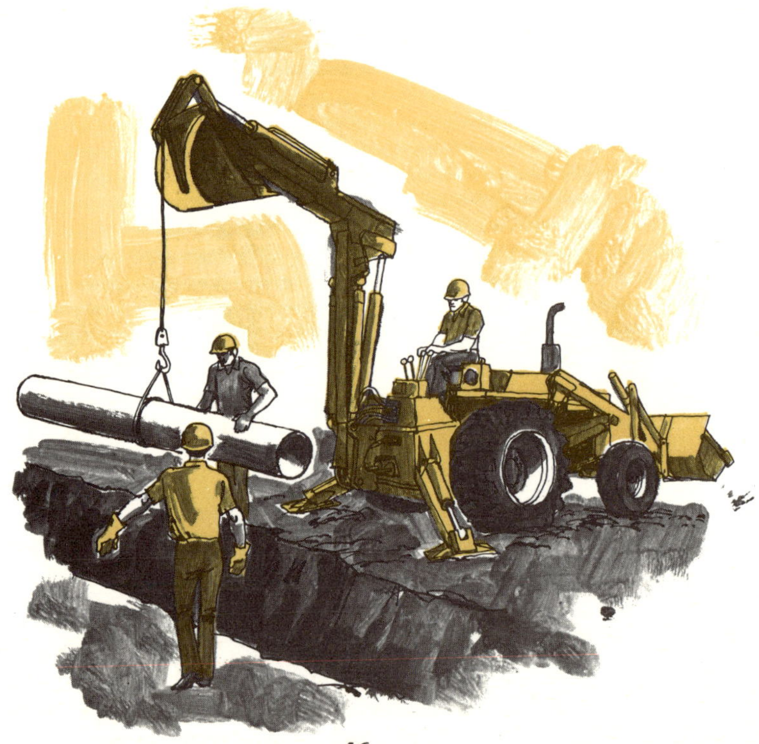

CITY AND HIGHWAY EQUIPMENT OPERATORS

Cities and states hire workers to operate heavy duty equipment used in *maintenance* of cities and highways.

You have probably seen the **STREET SWEEPER** at work. This man drives a machine that has large rotating brushes close to the ground. As the driver guides the machine, the brushes sweep up dirt and trash.

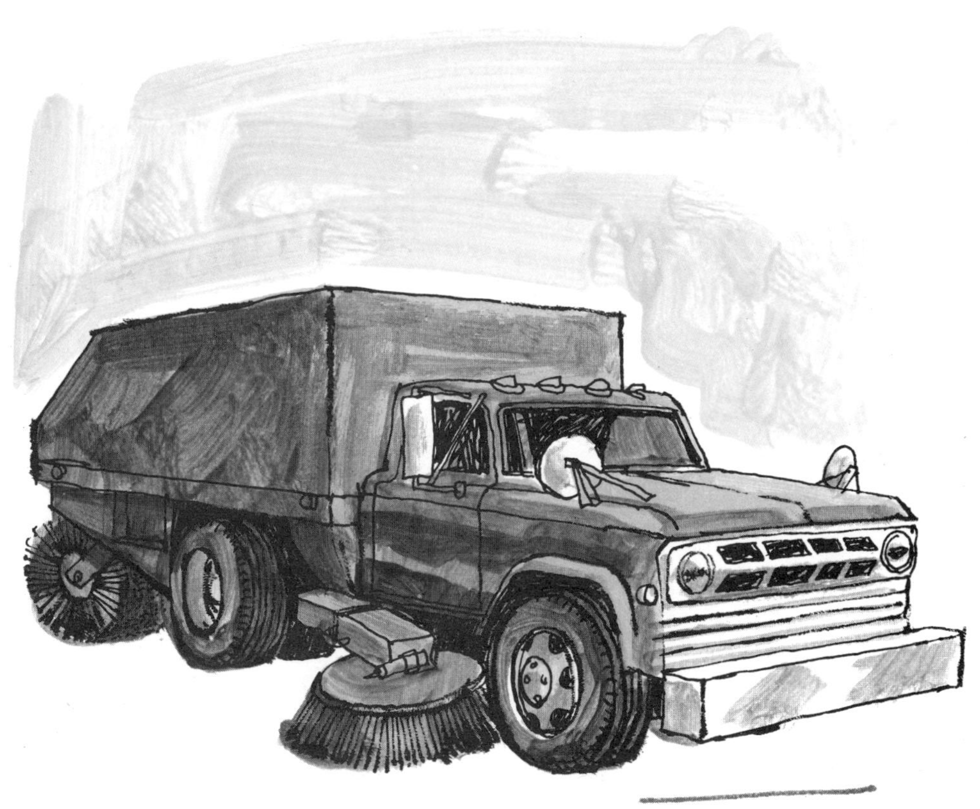

Both city and highway maintenance departments use a machine that the workers call the "giraffe." It is a tree-trimmer crane with a small platform at the end of a long mechanical neck. This folding neck is bolted to the bed of a truck.

A worker rides high into the air. From there he can trim branches from trees, repair signs, and rescue cats. At Christmas time he hangs the city's decorations high above the street.

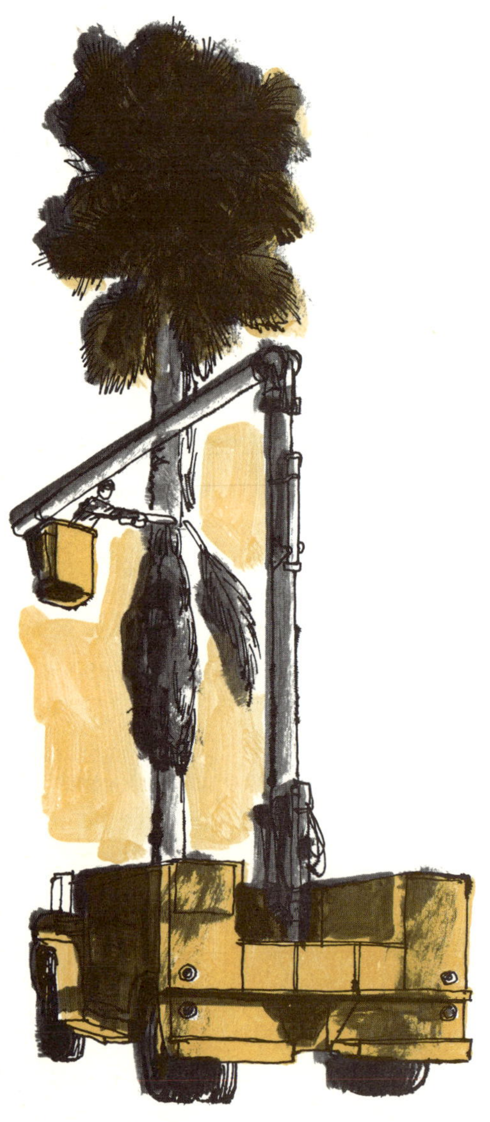

The city or highway maintenance department hires the drivers and the mechanics who repair the machines.

All the city's vehicles are kept in a large yard which has its own garage and mechanics.

If you are interested in driving equipment for your city, you might start out as a **FLAGMAN** helping control traffic when streets are being repaired. You might clean up machinery or fill the tank trucks used to water trees.

Young people who are dependable in the easier jobs can learn from their supervisors how to operate the maintenance equipment.

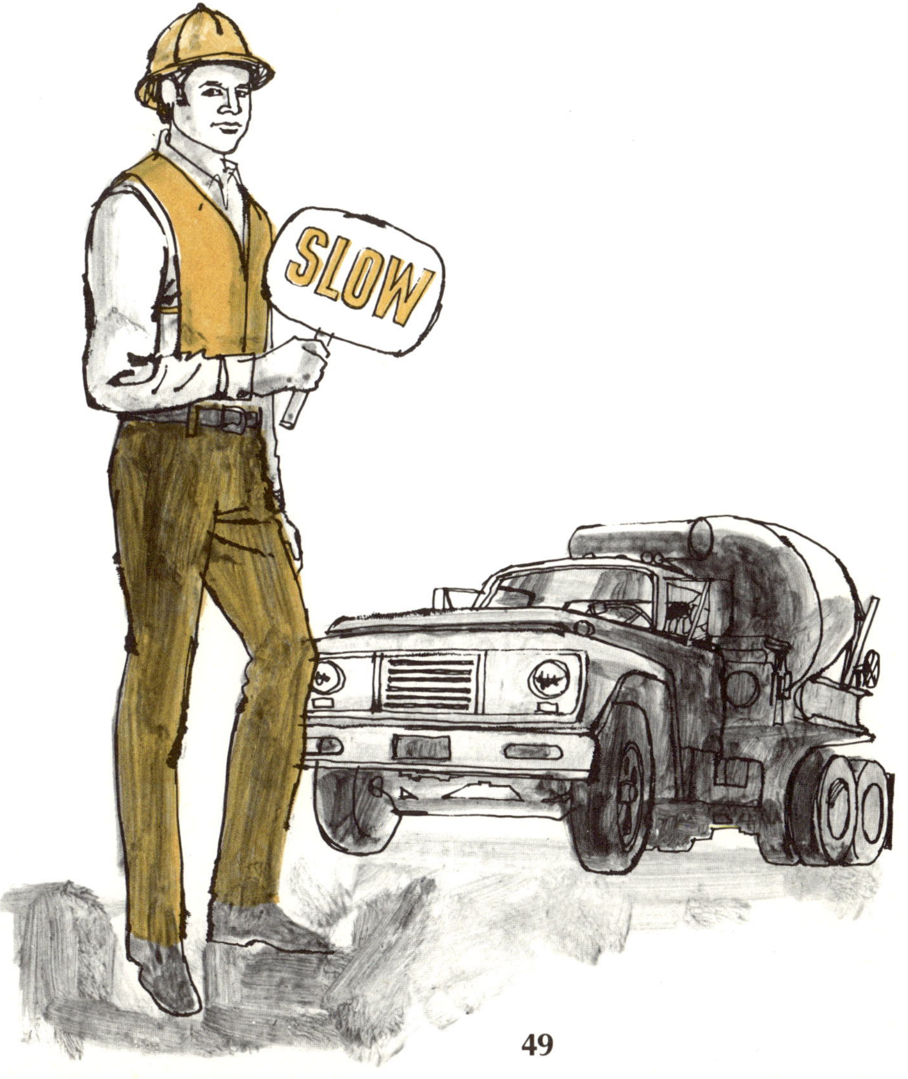

LANDSCAPING EQUIPMENT OPERATORS

Public parks and banks beside freeways and highways need to be planted and kept beautiful.

This gives many jobs to V.I.P. who operate *landscaping vehicles.* Many of these machines are much like farm vehicles.

A special, grass-planting machine is called a *hydro-seeder.*

It's a large tank truck. As its driver guides it slowly, a crewman, holding a large hose, sprays a mixture of water, grass seed, *mulch,* and fertilizer onto the soil. This way, the hydro-seeder crew can plant grass on slopes and in large park areas much more quickly than they could plant by hand.

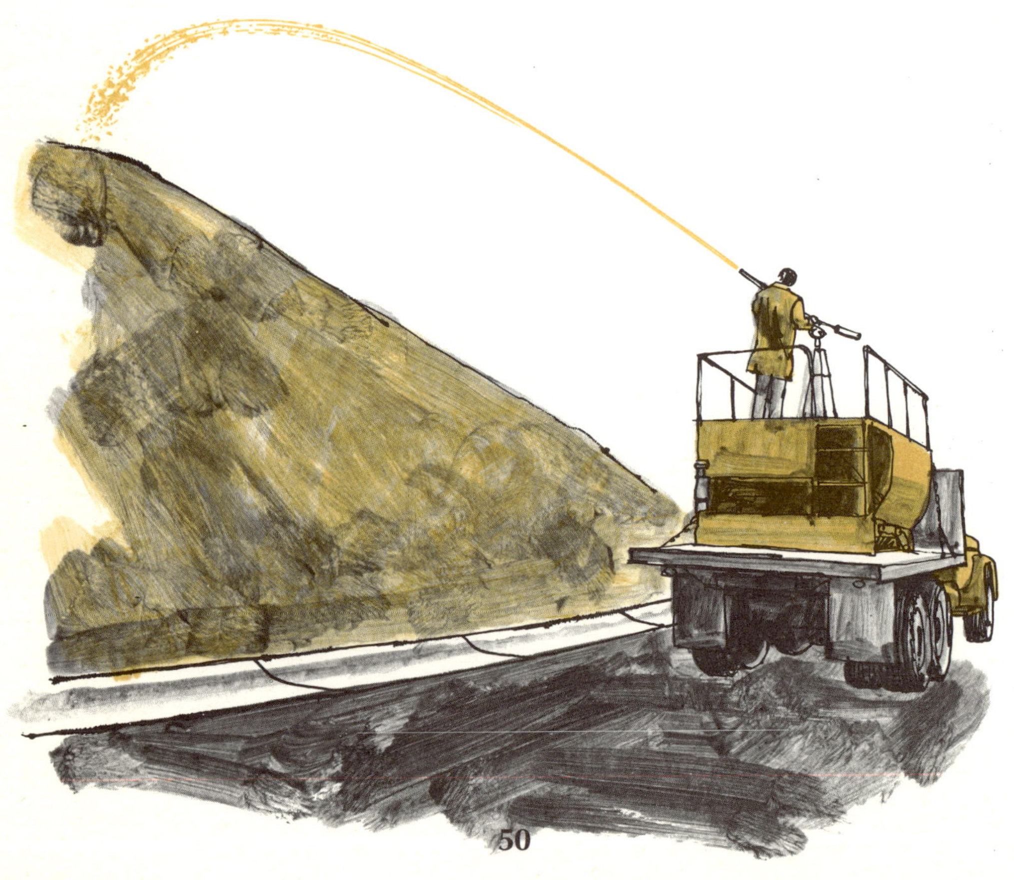

You may have seen workers riding across big lawns, driving *self-propelled lawn mowers.*

School grounds, parks, and golf courses need workers like these.

Many young people who like to work outdoors get jobs operating the power mowers. In warm climates these jobs may be found all year.

Beginners can learn on the job. Mechanical skills are helpful, but no special education is required.

Those who like the work may continue in the landscaping business. They can learn how to run the more complicated planting, digging, and cultivating machines.

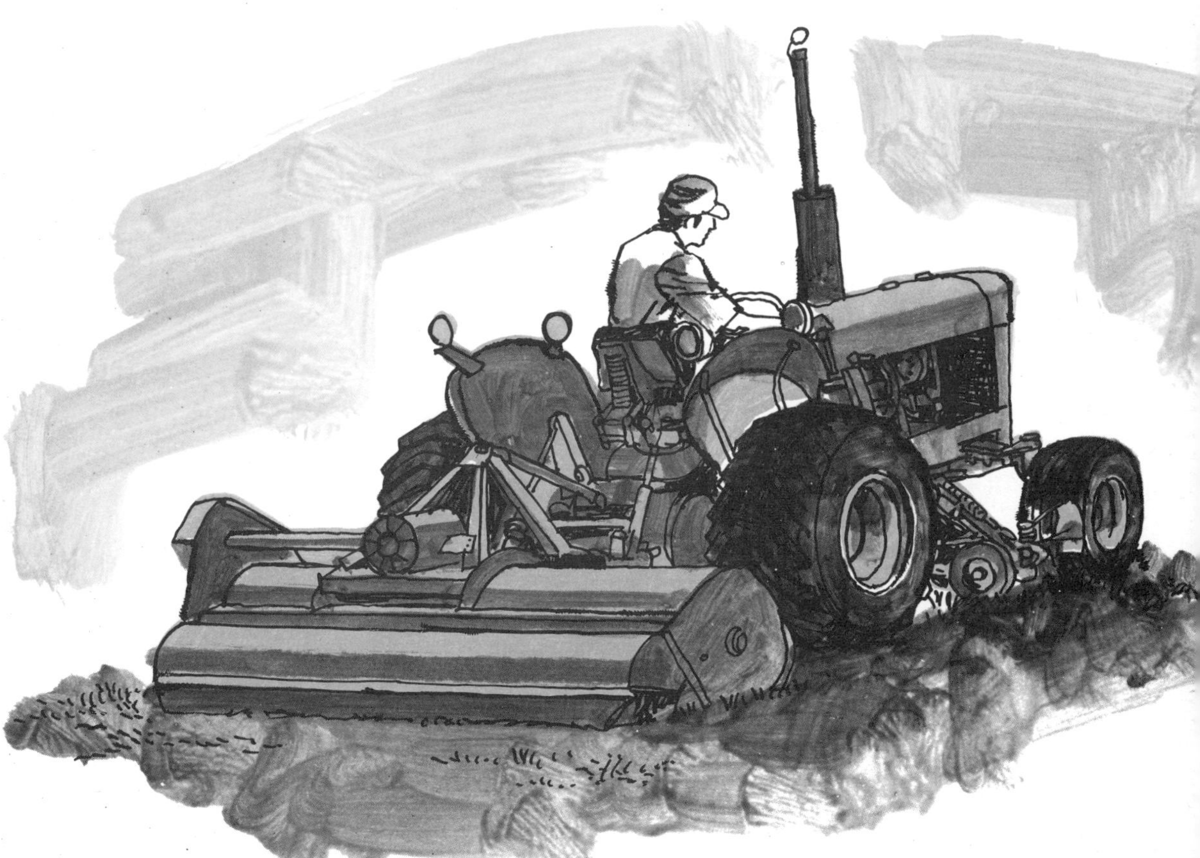

V.I.P. WHO SELL FARM AND EARTH-MOVING MACHINES

Each company that manufactures farm and earth-moving machines has stores, called *agencies* or *dealerships*. These are located in areas where the vehicles are most needed.

In the country, dealers may sell both farm and earth-moving equipment. In the cities, dealers usually sell equipment used in construction and road building.

Each dealer has several salesmen called **SALES ENGINEERS.** The sales engineers know a lot about construction and farming. They understand the job the customer wants the machine to do. Often the sales engineer advises the farmer or the earth-moving contractor on which machine will do the best job for him.

The sales engineer has large, colorful booklets that show pictures of all the equipment the company makes. These booklets also give a lot of information about the jobs the vehicles can do.

Salesmen at the agencies will give copies to anyone wanting to know about operating these machines.

The farm equipment salesman often goes out to the farms to talk to his customers. A farmer may need machinery, but he's often too busy to come to the agency.

The salesman may also go to the farm to make repairs and adjustments on the machinery he has already sold.

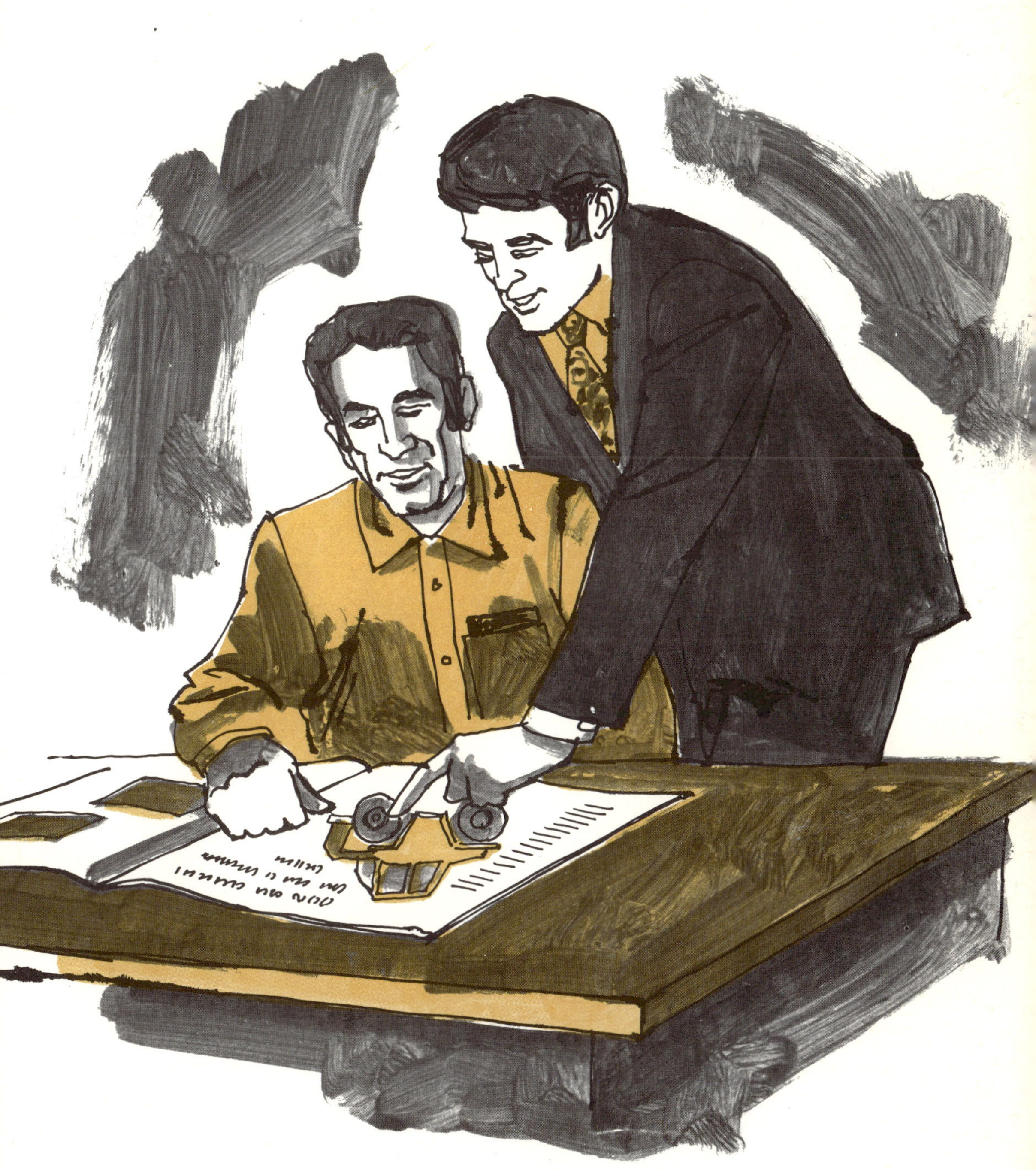

The construction equipment salesman keeps a list of all the *construction contractors* in his area who might need new equipment. He keeps track of the kinds of jobs they will be doing. When he goes out to the construction site, he knows whether the workers will be laying pipe, clearing land or paving a road. That way he can suggest the right equipment for the jobs and perhaps sell a machine to the contractor.

What kind of person makes a good salesman of farm or construction equipment? One who

- is able to get along well with people and enjoys helping them
- understands his customers' needs and problems
- is willing to travel to where his customers are
- knows a lot about the equipment he sells
- is able to fill out forms correctly
- has a high school or equivalent education

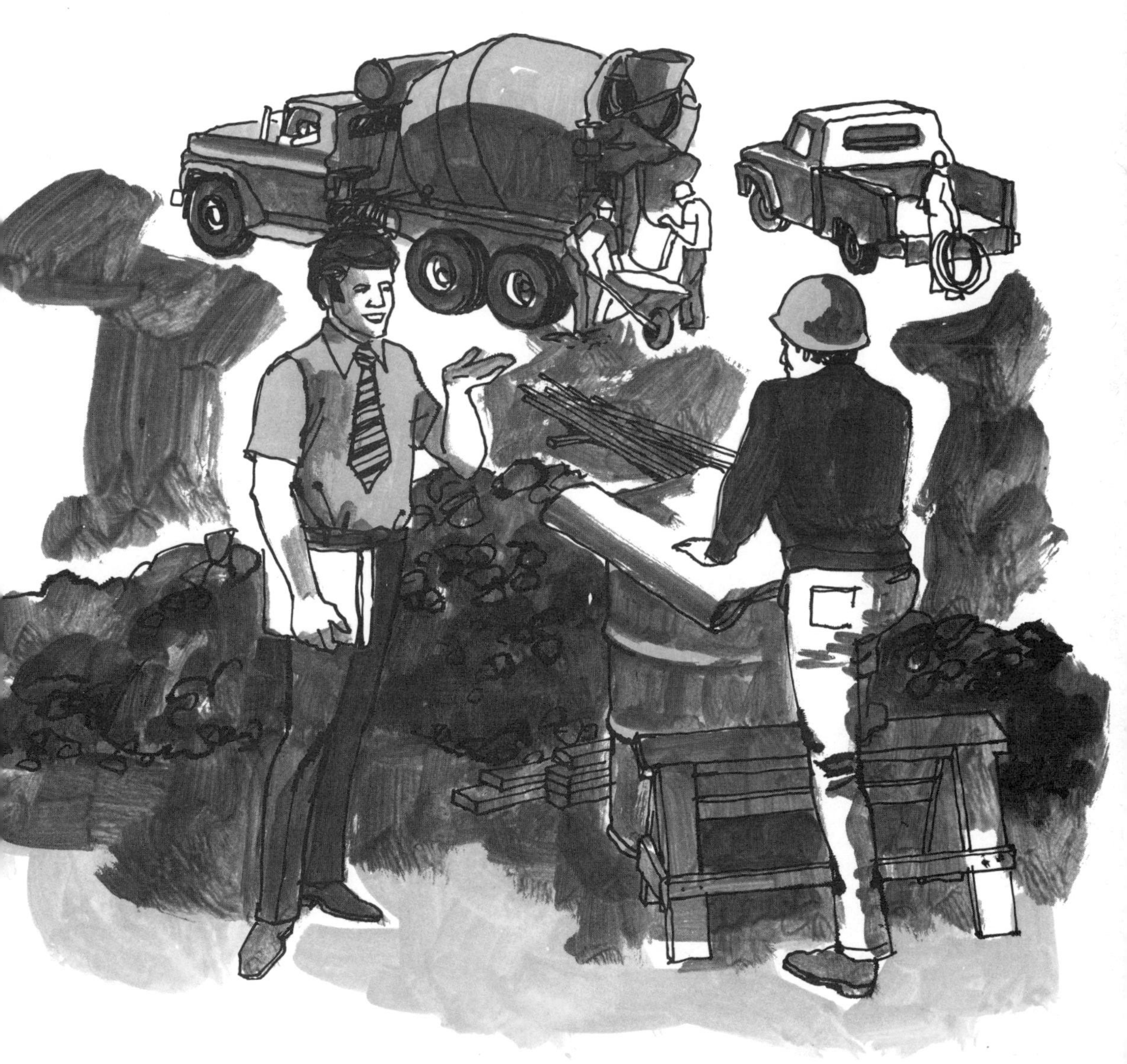

MACHINES IN YOUR FUTURE

Farm equipment operators and earth-moving equipment operators will continue to work with new and better machines.

The workers who can operate and repair the most complicated machines will be the most needed. They'll get the highest wages.

Scientists say that in the future, robot machines will cultivate, plant, and harvest the crops. A human operator will push buttons to direct the robots.

Perhaps you will operate farm machinery by these *remote controls*.

You may be working with under-water excavating machinery when *oceanographers* and construction engineers work together to build tunnels under the sea.

You may operate construction equipment on the moon, building shelters for explorers.

Today's jobs with farm and earth-moving equipment are very exciting. The jobs of the future will be even more exciting!

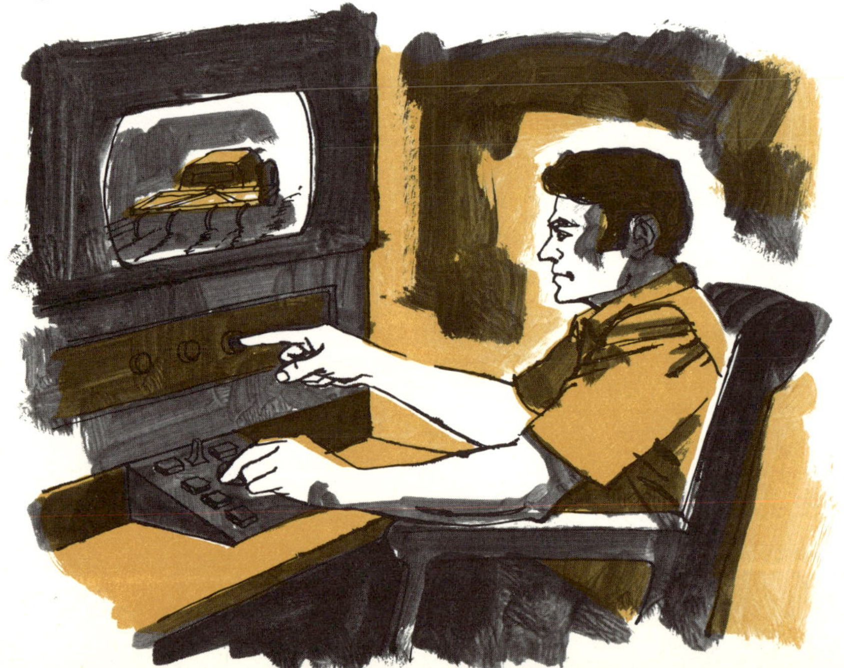

GLOSSARY/INDEX

Note Some of the words defined in this glossary have several meanings. The meanings described here are the ones used in this book.

**platform
attachment,**   **24**   a platform attached to the loadline of a crane that can lift workers or materials high above the ground

plow,   **14**   a tractor attachment that cuts grooves, or furrows, in the earth

power shovel   **23**
**power shovel
operator,**   **27**   a scoop at the end of a long mechanical arm attached to a tractor—It is run by a power shovel operator from controls in the cab of the tractor.

**pre-apprenticeship
training,**   **44**   practice in operating heavy equipment before a person begins an apprenticeship program

pyrometer,   **12**   a gage that measures high temperatures by checking the color of the heated material

remote controls,   **56**   electrical or electronic instruments that can control machinery from a distance

roller   **30**   a machine that packs and flattens a surface

roller operator,   **32**   a worker who drives heavy rollers that pack down the soil

sales engineer,   **52**   a person who sells farm and earth-moving equipment

scraper,   **30**
scraper operator,   **34**   a wide scoop pulled by a tractor which is driven by a scraper operator

**self-propelled
lawn mower,**   **51**   a lawn mower powered by its own electric motor or gasoline engine and operated by a driver